Oct

Steph —

Your legacy is something
I would hope to be
able to read.

May this book
inspire you on your
path.

Kurt

THE LEGACY

DAVID SUZUKI

THE LEGACY

AN ELDER'S VISION FOR OUR

SUSTAINABLE FUTURE

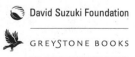

David Suzuki Foundation

GREYSTONE BOOKS

D&M PUBLISHERS INC.

Vancouver/Toronto/Berkeley

Greystone Books
An imprint of D&M Publishers Inc.
2323 Quebec Street, Suite 201
Vancouver BC Canada V5T 4S7
www.greystonebooks.com

David Suzuki Foundation
219-2211 West 4th Avenue
Vancouver, BC Canada V6K 4S2

Cataloguing data available from Library and Archives Canada
ISBN 978-1-55365-570-1 (cloth)
ISBN 978-1-55365-645-6 (ebook)

Editing by Nancy Flight
Copyediting by Barbara Czarnecki
Jacket and text design by Jessica Sullivan
Jacket photograph by Dorling Kindersley/Getty Images
Printed and bound in Canada by Friesens
Text printed on acid-free, FSC-certified, 100% post-consumer paper
Distributed in the U.S. by Publishers Group West

We gratefully acknowledge the financial support of the Canada Council
for the Arts, the British Columbia Arts Council, the Province of British
Columbia through the Book Publishing Tax Credit, and the Government of
Canada through the Canada Book Fund for our publishing activities.

Mixed Sources
Cert no. SW-COC-001271
© 1996 FSC
FSC

Contents

Foreword by Margaret Atwood *vii*

—[∽]—

Foreword

MARGARET ATWOOD

WHAT YOU'RE about to read is David Suzuki's Legacy Lecture. The term "legacy" has an ominous ring to it, a hint of departure: surely he isn't going somewhere? Not so soon! He's a landmark! No other living Canadian has done so much—nationally and internationally—to make us aware of the world we live in and of its precarious state. And no one else started this task so early and has taken so much flak for it.

It seems that David Suzuki has always been with us. He's lived in the tradition of the great prophets— those whose messages go unheeded because they tell us things we find uncomfortable. Time after time he's gone up the sacred mountain, listened to the voice, understood that it is what it is, and brought the hard but true words back down, only to find us cavorting around shiny gods of our own devising. He's been doing that in so many ways, over so many days—on *Quirks and Quarks,* a radio science program he started; on CBC television's *The Nature of Things,* which he's hosted since 1972; and through the David Suzuki Foundation, dedicated to making the world a sustainable place. It's a wonder he never gave up on us. But he didn't: after each potato flung his way, he trudged up the mountain again, rearranged the words to make them more understandable, and gave us another try.

As for his somewhat dire reputation—"Dr. Doom and Gloom," as he himself tells us—let's consider the deeper meaning of the word "legacy." A legacy is something you pass on, and it assumes there will be someone to pass it on to. That's quite a leap of faith for Dr. Suzuki, considering the grisly facts he's been facing. But as you'll see, he makes the leap. Human intelligence and foresight got us into our present pickle by enabling us to invent such efficient ways of exploiting Nature that our population growth went into overdrive, and now human intelligence and foresight are all we can rely on to see us through the tight bottleneck we're fast approaching—that narrowing chasm where far too many people are faced with far too little food and, very possibly, far too little air.

But, says Suzuki, we can do it if we really try, and we really will try if we can visualize the danger we're in. Programmed as we are to grasp the low-hanging fruit, enjoy the present hour at the expense of the years to come, and ignore the storm until it's almost upon us, we do have the capacity to learn from experience and to look ahead.

David Suzuki is by training a biologist—a scientist—which to some people conjures up the image of a white-coated rationalist, devoid of emotion and bent on pure experiment. But no human being is really like that, not even economists. Neurologists tell us that purely rational thinking is an impossibility for us: instead we think-feel; we feel-think. David Suzuki came to biology the way so many have: through the emotions, a love of the natural world—the world he then set out to explore using his intelligence. What he did with the love and the intelligence is a thing the human race has been doing to its advantage ever since the Pleistocene: he told stories about what he loved and what he discovered, stories that confer a benefit on those who hear them if only they will listen with care.

The "legacy" in this lecture is one of truthful words about the hard place we're in, but it's also one of hopeful words: our chance—if we will take it—for "opportunity, beauty, wonder, and companionship with the rest of creation." My own hope is that we ourselves will emulate David Suzuki and leave legacies in our turn, and that the planet will through our efforts become a better and more liveable home than the rapidly deteriorating biosphere we find ourselves in right now. It's the nature of gifts to pass from hand to hand; we should thank Dr. Suzuki for the gifts he has given and find within ourselves the grace to pass them on.

THE LEGACY

INTRODUCTION

Now that I'm in my seventies, I know that I am in the last part of my life—I call it the Death Zone—and that each day is a gift to be celebrated. Impending death is also a powerful motivation to reflect on life and the successes and failures, loves and losses, joys and tragedies, and people, experiences, and events that have shaped who I am—my values and beliefs.

Upon reaching retirement, university professors often deliver a last lecture, in which they pass on the accumulated wisdom of a lifetime. I have already had the temerity to write not one but two autobiographies, one when I reached fifty *(Metamorphosis: Stages in a Life)* and a second twenty years later *(David Suzuki: The Autobiography).* But they were recordings of my memories as best as I could recall, pieced together in a chronological sequence, but not deeply reflective.

I have recorded my thoughts and ideas extensively on subjects such as parenthood, academia, science, and politics, but always in bits and pieces in the hundreds of essays I've written as a journalist and columnist in newspapers, magazines, and journals. I have selected what I considered some of the best in collections such as *Inventing the Future, Time to Change, Earth Time, The David Suzuki Reader,* and *The Big Picture.*

But if I were to give a last lecture, what would I say? This book is based on the Legacy Lecture I delivered in December 2009 at the University of British Columbia, where I had been a professor for thirty-nine years. It is my version of a last lecture and an attempt to answer the age-old questions: What's it all about? What have I learned over a lifetime that I'd like to pass on? I also ponder how we as a species have arrived where we are today. We are so dazzled by our own inventiveness that we are blinded to the consequences of technology. We have very suddenly become a major planetary force and have discarded traditional perspectives, believing that what's most recent is the best. Finally, I offer my vision, based on a lifetime's worth of experiences, for a future that is possible, one rich in joy, happiness, and meaning.

Only by confronting the enormity and unsustainability of our impact on the biosphere will we take the search for alternative ways to live as seriously as we must. As an elder, I am impelled by a sense of urgency that comes from the recognition that my generation has induced change and created problems that we bequeath to my children and grandchildren and all generations to come. That is not right, but I believe that it is not too late to take another path.

Evolution of a

SUPERSPECIES

VER SINCE our species appeared on Earth, human beings have gathered around a fire to fulfill our most elemental human need—companionship— as we reaffirm kinship and tribal bonds, recount experiences, share insights, and ponder the great questions that have troubled us for so long:

Who are we?

How did we get here?

Why are we here?

Where are we heading?

Our answers to those questions are profoundly rooted in place—think of the Inuit in the Arctic, the San of the Kalahari Desert, the Aboriginal people of

Australia, and indigenous populations from coastal rainforests to prairies to mountainsides throughout the Americas. Over millennia, countless stories, songs, and dreams have expressed the multitude of ways the human mind has imagined the world into existence.

In the distant past, accumulated experience, observation, and insight shaped the answers to our ancient questions and thus created the way we perceived the world. According to the Haida of the Northwest Coast of North America, Raven picked up a clamshell and dropped it on a beach in Haida Gwaii, and from it emerged the first human beings. The Norse creation myth tells of the giant Ymir, whose body became the land; his blood became the oceans, his bones the mountains, and his hair the forests. Obviously, creation myths are not meant to be taken as literal history. The real meaning of or lessons within our creation myths are often buried in layers of elaboration, superstition, and metaphor and therefore may seem too incredible to be taken as more than fantastic stories.

Humans gather together and learn the meaning of the universe, our cosmology.—BRIAN SWIMME, cosmologist

I am a man of science, and today science is the source of the powerful insights that have become part of the modern narrative of how we got here. But the creation stories emerging from that science seem as far from our everyday experience and every bit as fantastic as myths of the past, and the real significance for us must be teased out of the arcane language.

Imagine a beginning 14 billion years ago in which the entire universe was contained within a single point the size of a period on a typed page. Consider the Big Bang, an explosion at such a high temperature that matter could not exist. As immense clouds of swirling gases cooled in the expanding universe, however, they condensed into particles of matter that thenceforth defined the basic physical rules throughout the cosmos. It confounds our imagination to think that vast clouds of atoms, drawn together by gravitational pull, eventually coalesced into countless stars that suddenly ignited their nuclear furnaces to light up the heavens in a cosmic instant.

And the scientific depiction of the evolution of life on Earth is inspirational, a saga of resilience and adaptability: after hundreds of millions of years on a sterile planet, one cell arose in the ocean that out-competed all others to become the mother of all life on Earth. Her descendants invaded every nook and cranny of the planet, as they transformed themselves into countless species in an ever-changing environment.

Early in the history of life, Nature began to shape new species to fit into habitats already occupied by other species. Never since the Archaean Period has a living thing evolved alone.—VICTOR B. SCHEFFER, zoologist

Traditional Narratives

FOR MOST OF human existence, we were oral creatures, sharing experiences, insights, and beliefs through the stories we told. Woven into the stories were implicit lessons about how to respond to the world around us. As Ivaluardjuk, an Inuit hunter, recounted to the Arctic explorer Knud Rasmussen in 1930, "The greatest peril of life lies in the fact that human food consists entirely of souls. All the creatures that we have to kill and eat, all those that we have to strike down and destroy to make clothes for ourselves, have souls like we have, souls that do not perish with the body, and which must therefore be propitiated lest they should revenge themselves on us for taking away their bodies."

Until very recently, all of humankind was local and tribal, following game and plants through the seasons in a nomadic existence of hunting and gathering. Contrary to popular belief, our ancient ancestors were not slow-witted, primitive savages. They were human beings with the same genetic heritage as twenty-first-century people. As the late giant in anthropology Claude Lévi-Strauss pointed out, "I see no reason why mankind should have waited until recent times to produce minds of the calibre of a Plato or an

Einstein. Already over two or three hundred thousand years ago, there were probably men of similar capacity who were of course not applying their intelligence to the solution of the same problems as these more recent thinkers."

To Lévi-Strauss, scientist and shaman approach the world from very different perspectives, but both perspectives provide profound insights, neither invalidating the other: "Certainly the properties to which the savage [or Native] mind has access are not the same as those which have commanded the attention of scientists. The physical world is approached from opposite ends in the two cases: one is supremely concrete, the other supremely abstract; one proceeds from the angle of sensible qualities and the other from that of formal properties."

Human beings appeared on the plains of Africa 150,000 to 400,000 years ago, when woolly mammoths, sabre-toothed tigers, and giant sloths still flourished and the savannahs of Africa were filled with animals in numbers and variety beyond anything we know today. There was little in the appearance of our distant ancestors to suggest the explosive change we would undergo as we left our African birthplace to populate every part of the globe in a mere 150 millennia. I doubt that any other species would have trembled at the sight of our ancestors and whispered, "Watch out for those two-legged, furless apes. They're going to take over the planet."

We were not an impressive species in numbers, size, speed (an elephant can outrun the fastest human on Earth), strength (a chimpanzee weighing a hundred pounds could whip me and probably you too), or sensory acuity (I know if I were swinging through the trees on a vine, without glasses, I'd smack into a tree and probably fall to the ground to be eaten by a sabre-toothed tiger). The secret to our success was invisible, the two-kilogram organ encased in our skulls.

The human brain conferred a massive memory, insatiable curiosity, and remarkable creativity, qualities that more than compensated for our lack of physical or sensory abilities. And that brain had become aware of itself, conscious of its presence in time and space, capable of imagination and dreams. We observed; learned from accidents, mistakes, trial and error, and discoveries; remembered what we had experienced; recognized causal relationships; and came up with innovative solutions to problems.

The brain evolved in a biocentric world, not a machine-regulated world. It would be therefore quite extraordinary to find that all learning rules related to that world have been erased in a few thousand years.—EDWARD O. WILSON, ecologist

Drawing on our experience and knowledge, we dreamed of our place in the world and imagined the future into being. By inventing a future, we could look ahead and see where dangers and opportunities lay and recognize that our actions would have consequences in that future. Foresight gave us a leg up and brought us into a position of dominance.

Our creation stories and origin myths provided answers to those eternal fireside questions. Distilled from generations of observation and insight, they were carefully nurtured and handed on to those who followed, providing meaning and insight into their lives.

Throughout human existence, elders have been the repository of experience, of knowledge painstakingly acquired over centuries about our origins, our purpose, and our destiny. And now I too am an elder. Within the span of my living memory, which encompasses stories from the lives of my grandparents that reach back to 1860, cataclysmic changes have taken place in society and the world.

Elders are not "senior citizens"... They are wisdom-keepers who have an ongoing responsibility for maintaining society's well-being and safeguarding the health of our ailing planet Earth.—ZALMAN SCHACTER-SHALOMI, rabbi, and RONALD S. MILLER, writer

I was born in 1936 in Vancouver, British Columbia, Canada, where both of my parents were born as well. My parents named me David Takayoshi Suzuki—David, a powerful name given to me by my father, who feared I'd be a small man surrounded by Caucasian Goliaths; Takayoshi, meaning filial piety—respect for elders—conferred on me by my father's father; and Suzuki—Bellwood in English—the biggest clan in Japan.

Both sets of my grandparents were born in Japan in the 1860s. The population of the world had reached 1 billion only a few decades earlier, passenger pigeons still darkened the skies, and Tasmanian tigers stalked the Australian landscape.

Canada was born when my grandparents were infants, Japan was casting aside almost three hundred years of feudalism of the Edo Period to embrace

Dad's catch on the Vedder River, 1940

Western industrialization in the Meiji Restoration, and virtually every technology that we take for granted today—from telephones to cars, plastics, antibiotics, and computers—was still to be invented.

My mother's parents
newly arrived in Canada

[OUR RELATIONSHIP WITH THE PLANET]
Within my living memory, the human relationship
with the planet has transmogrified—we have become
a force like no other species in the 3.8 billion years of
life's existence on Earth. And the ascension to this
position of power has occurred with explosive speed. It
took all of human existence to reach a population of 1
billion early in the nineteenth century. Since then, in
less than two centuries, it has shot past 6.8 billion.

Each time the population doubles, the number of
people alive is greater than the sum of all other people
who have ever lived, but now we are also living more
than twice as long as people did in the past. We are
the most numerous mammal on the planet, and our
numbers and longevity alone mean that our ecological
footprint is huge; it takes a lot of air, land, and water to
meet our basic needs.

In the twentieth century, cheap, portable, energy-
rich fossil fuels greatly increased our technological
capacity, further amplifying our ecological footprint.
While it once took months for First Nations people
to cut down a giant cedar tree, today one man and a
chainsaw can do it in minutes. Modern fish boats stay
at sea for weeks, loaded with radar, sonar, freezers, and
nets big enough to hold a fleet of 747s.

Ever since the end of World War II, we have enjoyed
and come to expect a never-ending outpouring of

consumer goods. Indeed, after the terrible shock of 9/11, Florida governor Jeb Bush, brother of the U.S. president, stated that it was Americans' "patriotic duty to go shopping." A global economy built on supplying that ever-expanding consumer demand exploits the entire planet as a source of raw materials.

As a result of our numbers, vast technological muscle power, exploding consumption, and global economy, the human footprint can be seen from a

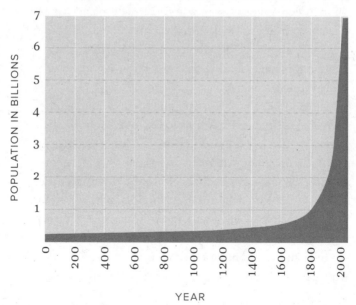

GLOBAL HUMAN POPULATION GROWTH

plane kilometres above the earth—in the huge lakes stretching behind dams, vast areas of clear-cut forests, massive farms, and immense cities criss-crossed by straight lines of roads, encased in a dome of haze, and ablaze with lights in the depth of night.

When man becomes greater than nature, nature, which gave him birth, will respond.—LOREN EISELEY, anthropologist

Throughout the history of life on Earth, living organisms have interacted with and altered the chemical and physical features of the planet—weathering rock and mountains, generating soil, filtering water as part of the hydrologic cycle, sequestering carbon as limestone, creating fossil fuels, removing carbon dioxide, and adding oxygen to the atmosphere. But those processes took millions of years and involved tens of thousands of species. Now one species—us—is single-handedly altering the biological, chemical, and physical properties of the planet in a mere instant of cosmic time.

We have become a force of nature, a superspecies; and it has happened suddenly, with explosive speed. Not long ago, hurricanes, tornadoes, floods, drought, forest fires, even earthquakes and volcanic explosions were accepted as "natural disasters" or "acts of God." But now we have joined God, powerful enough to influence these events.

Although our creative abilities have led to impressive technologies, they have often been accompanied by costs that could not be anticipated because our knowledge of how natural systems work is so primitive. When theoretical physicists discovered that splitting atoms would release vast amounts of energy, that process was harnessed to create the atomic bombs that exploded over Hiroshima and Nagasaki and brought World War II to a quick end.

But when the bombs were detonated in 1945, scientists didn't know about radioactive fallout, which was discovered only when bombs were exploded on the Bikini atoll years later. More years passed before scientists discovered that explosions set off electromagnetic pulses of gamma rays that knock out electrical circuits, and years after that, scientists realized that debris thrown into the atmosphere from atomic explosions might block the sun's rays and induce nuclear fall or winter.

When Paul Müller discovered that DDT kills insects, his employer, Geigy, recognized the potential for profit by using it as an insecticide. It appeared that the molecule had little biological effect on organisms other than insects, and Müller won a Nobel Prize for his discovery in 1948. As the use of DDT was vastly expanded over huge areas, birdwatchers began to notice a decline in bird populations, especially raptors such as hawks and eagles. Eventually, biologists tracked down the cause: as a result of the high concentration of DDT in the fatty shell glands of the birds, eggshells were becoming thinner and often broke in the nest.

Then extremely high levels of DDT were found in the breasts and milk of women. The explanation was a phenomenon called biomagnification, or the increase in concentration of a substance as it moves up the food chain. Thus, while an insecticide might be sprayed at low concentrations, micro-organisms absorbed and concentrated the molecules. When the micro-organisms were eaten by larger organisms, concentration was amplified again. So in top predators like raptors and humans, pesticide concentrations could be hundreds of thousands of times greater than when the pesticide was sprayed. No one predicted biomagnification, because it was discovered as a biological process only when birds began to disappear and scientists tracked down the cause.

And so it has gone as we apply new technologies, only to find later that they have unanticipated side effects. New technologies unleash perturbations that become a kind of experiment within the biosphere, as we will no doubt learn with genetically modified organisms. Our ignorance is vast, but human beings have become such a powerful force that we are altering the life-support systems of the planet.

The Consequences
of Relentless
Population Growth

THE CURVE OF exponential growth is a familiar one (figure 1).
Initially, with a small population, the curve rises very slowly.
Eventually, it reaches a point where it inflects sharply
upward, and then it tails off again as limits are encountered.
Nothing in the world can rise steadily forever, and humanity
is now in an explosive period of increasing numbers in
which the curve is turning sharply upward.

According to "'Science Summit' on World Population: A
Joint Statement by 58 of the World's Scientific Academies,"
"It took hundreds of thousands of years for our species to
reach a population level of 10 million, only 10,000 years
ago. This number grew to 100 million people about 2,000
years ago and to 2.5 billion by 1950. Within less than the
span of a single lifetime, it has more than doubled to
5.5 billion in 1993 ... We face the prospect of a further
doubling of the population within the next half century."

The rapid increase in human numbers has coincided
with accelerating technological innovation, which has risen
even more steeply over the same period as population
growth. Technology has enabled us to keep up with the
escalating demands on the planet from more and more

people. But greater numbers of people plus consumptive demand have placed a huge burden on the biosphere.

Again, the joint statement by fifty-eight scientific academies notes that "in the last decade, food production from both land and sea has declined relative to population growth. The area of agricultural land has shrunk ... The availability of water is already a constraint in some countries. These are warnings that the earth is finite and that natural systems are being pushed ever closer to their limits."

We have always assumed that increasing wealth and technological innovation would enable populations to grow indefinitely. In industrialized nations where people are not having enough babies to replace themselves, there may not be a large enough workforce to support an increasingly aging population. Because there is a tight correlation between population increase and economic growth, countries like Canada are using immigration to keep the numbers and economy growing. But the ecological costs are not factored in.

As a result, says the world academies statement, "If current predictions of population growth prove accurate and patterns of human activity on the planet remain unchanged, science and technology may not be able to prevent irreversible degradation of the natural environment and continued poverty for much of the world."

[EARTH]

*Conservation is a state of harmony between men and
land. By land is meant all things on, over, or in the earth.
Harmony with land is like harmony with a friend: you
cannot cherish his right hand and chop off his left. That
is to say, you cannot love game and hate predators: you
cannot conserve the waters and waste the ranges; you
cannot build the forest and mine the farm. The land is
one organism.*—ALDO LEOPOLD, ecologist

The biosphere is the layer of air, water, and land
where all species live. It is extremely thin. If Earth were
shrunk to the size of a basketball, the layer of topsoil on
which our food is grown would be a single atom thick.
And on that thin organic mix, humanity's survival rests.

Soil is a creation of life, as dead and decaying micro-
organisms, animals, and plants are added to the matrix
of clay, sand, and gravel. Along the central plains of
North America, soil was built from the annual contribution of leaves falling from deciduous forests, prairie
grasses, and droppings from vast populations of passenger pigeons and bison. The flood plains of northern
China, the confluence of the Tigris and Euphrates in
Mesopotamia, and the Nile River enabled the growth
of great civilizations. It takes centuries to create a centimetre of soil in the richest areas. But modern-day
farming practices are depleting in decades what took
nature tens of thousands of years to create.

We denigrate soil as "dirt," but it is a living community of organisms. It is a world that we barely know. In a single teaspoon of soil, we may find hundreds of millions to 3 billion bacteria and a million fungi, like yeast and moulds. There is a veritable zoo of creatures in soil, from microscopic fungi, bacteria, yeast, protozoa, rotifers, and roundworms to creatures on the edge of visibility, such as mites and springtails, to the larger woodlice, earthworms, beetles, centipedes, slugs, snails, and ants, and finally to the giants, including moles, rabbits, and other rodents.

The different groups perform services that keep soil alive. Bacteria and fungi decompose matter into detritus, which earthworms ingest and excrete as soil nutrient. Worms rummage through massive amounts of soil, enabling water, air, and organic material to percolate into the matrix.

Dryland ecosystems, which are defined by the amount of annual rainfall they receive, range from arid (less than 200 millimetres of winter rain) to semi-arid (200 to 500 millimetres winter rain) and hyper-arid. These areas receive so little precipitation that the evaporation rate is far greater than the amount that rains. Nevertheless, dryland ecosystems provide crops, fuelwood, and livestock, but they are particularly vulnerable to degradation into desert. In 2000, a third of all people lived in drylands, 10 to 20 per cent of which were already degraded (that's 6 to 12 million square kilometres). Desertification is occurring in 70 per cent of drylands.

Each year, an estimated 24 billion tons of topsoil is lost throughout the world, in large part because of agricultural practices and desertification. The amount lost over two decades is equal to the entire cropland of the United States. Annually, that represents a lost productivity worth over $40 billion.

[AIR]

Beyond the air there is only emptiness, coldness, darkness. The "boundless" blue sky, the ocean which gives us breath and protects us from the endless black and death, is but an infinitesimally thin film. How dangerous it is to threaten even the smallest part of this gossamer covering, this conserver of life.—VLADIMIR SHATALOV, Soviet cosmonaut

The atmosphere supports all terrestrial organisms with life-giving oxygen, is the source of weather and climate, and is a critical part of the hydrologic cycle. It is easy to think of the atmosphere as reaching all the way to the stars, and while it does extend 2,400 kilometres away from Earth's surface, 99 per cent of its mass is within 30 kilometres and 50 per cent is within 6 kilometres. The first 11 kilometres is called the troposphere, the zone where all life is found and weather occurs.

Oxygen is a highly reactive element, the agent of oxidation, or rust, so before life appeared on Earth, chemical reactions would have scrubbed any oxygen from the air. The primordial atmosphere before life appeared is thought to have been made up of free hydrogen, carbon dioxide, water vapour, and possibly methane and ammonia, air that would be toxic to animals like us.

It was photosynthesis that altered the balance of air's ingredients, as plants took up carbon dioxide to construct the basic molecules of life's architecture. The energy in photons from the sun was transformed into chemical energy that could be used when needed. In return, oxygen was liberated and over millions of years transformed the atmosphere into what it is today.

Now we are altering its chemistry by spraying vast quantities of chemicals from planes and emitting gases from our chimneys, exhausts, and vents. All of humanity is burning fossil fuels at unprecedented rates and adding greenhouse gases in quantities greater than at any other time in all of human existence. Greenhouse gases such as carbon dioxide, methane, and water vapour occur naturally, while more potent molecules like CFCs are synthesized by us. They allow sunlight to penetrate the atmosphere but, like a blanket, reflect infrared (heat) wavelengths back onto Earth's surface. The continued addition of greenhouse gases to the atmosphere, thereby thickening the blanket effect, is projected to have potentially catastrophic ecological consequences.

In nature, everything is connected, so as more heat-trapping gases are added to the atmosphere, polar ice sheets begin to melt, ocean waters warm and expand, and terrestrial ecosystems begin to change as animals and plants move in order to remain within their temperature comfort zone.

[FIRE]

The sun is the primary source of energy for the enormous web of living things. Through photosynthesis, plants capture sunlight and convert photons of energy into stable molecules of sugar, where the energy is tied up in chemical bonds. In a miraculous bit of biological alchemy, plants inhale carbon dioxide from the atmosphere, add water from the soil, and, with energy from sunlight, create chains and rings of carbon that are the backbones of all the large carbon-based molecules of life. Just as our car gas tanks store fossilized sunlight in oil, plants can store sunlight as chemical energy in sugars for use later when needed. Animals can also parasitize that stored sunlight by eating plants (herbivores) or eating animals that eat the plants (carnivores). Thus, the web of living things is built on photosynthesis to exploit the sun's gift of energy. In any organism, oxygen "burns" the sugar by breaking the chemical bonds between carbons, sticking on oxygen to make carbon dioxide, and liberates sunlight's energy. We are the sun, and this allows us to metabolize, move, grow, and reproduce.

Humanity has co-opted, or taken over, more than 40 per cent of the photosynthetic activity on the planet. We may replace wild organisms on land with crops grown for our own use or for our domestic animals or remove the photosynthetic potential by flooding, burning, or developing areas. As we destroy or exploit that photosynthetic energy, other species are deprived of it. Since we are a single species out of some 10 to 30 million species, we have clearly become bloated beyond all balance. The challenge and opportunity will be for us to allow much more of the photosynthetic activity to return for use by the rest of life while we find other ways of recovering energy through photovoltaics, windmills, tide power, geothermal energy, and so on.

We now know that photosynthesis is a critical factor in the removal of carbon dioxide from and contribution of oxygen to the atmosphere. So long as the carbon-based molecules created by photosynthesis are stored in the structures of trees and other plants, then that carbon remains bound up, or sequestered, and out of the atmosphere. Thus, the protection and preservation of photosynthetic activity must be an important consideration as we try to minimize the impact of climate change.

[WATER]

Oceans cover 71 per cent of Earth's surface. And their water flows like a giant conveyor belt in great currents around the planet and helps stabilize planetary temperatures by absorbing heat in equatorial areas and releasing it in the northern or southern regions. In polar regions, supercooled water sinks deep into the ocean and then flows very slowly, cooling the surrounding depths as it moves. Like forests on land, marine plants absorb and sequester about half of the carbon dioxide taken up by living organisms and harbour much of Earth's biodiversity. And the diversity of living organisms has created numerous cultures based on those creatures. Japanese culture would be radically different without seafood, as would that of numerous First Nations along North America's coasts.

When I was in high school in the early 1950s, a teacher told us that the oceans teem with life in such abundance that we couldn't catch them fast enough. (It almost seemed like we had to catch more and more or the fish would take over the ocean.) Those fish provide limitless protein, she said. That may have been true in the '50s, but it is certainly not today. The human population has more than doubled since then, and that fact alone has escalated demand for fish and dramatically reduced fish populations. But our powerful new technology and our ability to take vast quantities of fish in ever more remote areas have also contributed to that decline. Marine biologist Boris Worm has calculated that if we do not radically change our ways, by 2048 there will be no commercially viable fish species left in the oceans.

Today's global fishing fleet boasts large vessels equipped with technology that enables the boats to stay at sea through all kinds of weather for weeks at a time and range over immense distances. Fibres in fishing nets are so strong that two huge trawlers can drag heavy rollers along the bottom of the deep and trail nets that are big enough to hold several jumbo jets. In the quest for specific target species, fishers often discard more than twice as much fish as "bycatch," species that are either considered worthless or not allowed to be sold.

Until they were outlawed, tens of thousands of kilometres of drift nets held at the ocean's surface by floats captured animals indiscriminately—including squid, fish, turtles, birds, and mammals. The nets were curtains of death. Now banned, drift nets have been replaced by longlines, ropes that are kilometres long and hold thousands of hooks, which capture targeted fish as well as sharks, turtles, and birds.

The ocean surface is used as a highway for immense vessels, such as cruise ships carrying thousands of people and supertankers transporting oil. Deliberately or accidentally, they dump or spill waste and garbage into the water and introduce alien species transported in ballast water.

Entire cities also use the oceans as dumping grounds for raw sewage and chemical effluent, while runoff from farmland washes pesticides and fertilizers into the sea, creating "dead zones" that lose oxygen and become toxic to life. Dead zones are increasing in size, duration, and number in all the oceans of the world. It is symbolic of our thoughtlessness that the capitals of Canada's two coastal provinces, Halifax in Nova Scotia and Victoria in British Columbia, dump raw sewage into the Atlantic and Pacific Oceans, respectively.

Time... is growing short. Nature's machinery is being demolished at an accelerating rate, before humanity has even determined exactly how it works. Much of the damage is irreversible.—PAUL ERHLICH, ecologist

Plastic and garbage tossed into the ocean end up in huge oceanic gyres where currents flow in giant circles and centrifuge the material into immense islands of debris. Some islands are bigger than the state of Texas. The action of waves, water, and sunlight breaks down much of the plastic into nurdles, small, fingernail-sized bits that are perfect for animals like albatross or turtles, which are programmed to eat anything that size. Autopsies reveal sea animals' stomachs crammed with such plastic pieces.

The interface of the oceans' surface with air is the largest area on the planet. Evaporation and absorption mean that atoms and molecules flow back and forth at the interface. As we burn ever-greater quantities of fossil fuels, carbon dioxide levels are rising in the atmosphere and hence even greater quantities of CO_2 are entering the water, where it forms carbonic acid. Consequently, the oceans are becoming more acidic, and acidification interferes with the development of many forms of life that use calcium carbonate to build protective shells. That puts at risk such animals as oysters and crabs as well as many forms of plankton, which are the very basis of the marine food web.

[BIODIVERSITY]

The vast tapestry of plants, animals, and micro-organisms is the source of all we need to survive. Plants create our oxygen-rich atmosphere and remove carbon dioxide by photosynthesis, and the sun's energy captured in the process provides the energetic mole-cules that our bodies use to move, grow, and reproduce. Plant roots, fungi, and the micro-organisms in soil fil-ter water as it percolates through the earth. Every bit of the food that we require to nourish our bodies was once alive, while the carcasses of dead organisms release their macromolecules to create soil in which life grows.

Yet today we are tearing at this web of life, which is the source of our most fundamental needs, by driving an estimated fifty thousand species to extinction every year. It is humbling to realize that if our species were to go extinct overnight, biodiversity would rebound around the planet. In contrast, the loss of an insect group such as all ants would result in a catastrophic collapse of terrestrial ecosystems.

The current global extinction spasm threatens entire ecosystems as well as individual species, and the most vulnerable organisms are those predators such as whales, tigers, and grizzlies at the apex of the food chain. There is no predator higher on the food chain than us.

There is no escape from our interdependence with nature; we are woven into the closest relationship with the Earth, the sea, the air, the seasons, the animals and all the fruits of the Earth. What affects one affects all—we are part of a greater whole—the body of the

planet. We must respect, preserve, and love its manifold
expression if we hope to survive.—BERNARD CAMPBELL,
anthropologist

[CRISIS]

For most of human existence, we were local tribal animals who didn't have to worry whether there were
people across an ocean, over a mountain, or on the
other side of a forest or desert. We lived with what we
had in our local environment. Even with the simplest
of implements, people were able to take too much,
extirpating slow-moving, easily caught animals or cut-
ting and burning too many trees. Human migration
may have been motivated by ecological degradation as
much as curiosity or social conflict. Over time, those
left behind had to learn to live in balance with the surroundings of their own tribal territory.

We now draw borders around our territory, from
our own property to cities, provinces, and nations, and
we're prepared to fight and die to protect those boundaries. But now we exist in a different relationship with
Earth.

Today, for the first time, we have to ask, "What is the
collective impact of all 6.8 billion people in the world?"
As we continue to cling to the primacy of our borders, we fail to see that air, water, seeds, and soil that

blow across oceans and continents, or migrating fish, birds, mammals, and insects, pay no heed to human priorities.

In contrast to the way we view our surroundings, every aspect of the natural world is interconnected. Thus, for example, we fragment the world into areas like economics, energy, health, and environment and try to manage them separately. But the kind of energy we use has immense repercussions in the air, water, and land, which in turn have a huge impact on human health and thus the economy.

Our great crisis of climate change demands a radically new approach. We must act as a single species to deal with global problems that transcend borders and acknowledge the rules and limits that nature defines. We must also act in the best tradition of our species, which has been to use foresight based on our knowledge and experience. And today we have the amplified ability to look ahead with scientists armed with super-computers and global telecommunications.

In 1992, 1,700 senior scientists from seventy-one countries, including 104 Nobel Prize winners (more than half of all laureates alive at that time), signed a document called "World Scientists' Warning to Humanity."

Their opening words were an urgent call to action: "Human beings and the natural world are on a collision course. Human activities inflict harsh and often irreversible damage on the environment and on critical resources. If not checked, many of our current practices put at serious risk the future that we wish for human society... and may so alter the living world that it will be unable to sustain life in the manner that we know. Fundamental changes are urgent if we are to avoid the collision our present course will bring about."

They went on to list the areas of collision, from the atmosphere to water resources, oceans, soil, forests, species, and population; then their words grew even more dire: "No more than one or a few decades remain before the chance to avert the threats we now confront will be lost and the prospects for humanity immeasurably diminished. We the undersigned, senior members of the world's scientific community, hereby warn all humanity of what lies ahead. A great change in our stewardship of the earth and life on it, is required, if vast human misery is to be avoided and our global home on this planet is not to be irretrievably mutilated." The document concluded with a list of the most urgent actions to be started immediately.

This is a frightening document, especially when you consider that scientists of such stature do not readily sign such strongly worded statements. The failure of the media, politicians, and corporations to respond to their warning means we are turning our backs on the survival strategy of our species—look ahead, identify the dangers and opportunities, then act accordingly to avoid the hazards and exploit the opportunities. The crisis is real, and it is upon us.

Finding a
NEW PATH

AN WE marshal the vision and energy to change direction and follow a new path? We have no choice. The Chinese symbol for crisis is made up of two parts: danger and opportunity.

danger  *opportunity*

We have discussed the danger. The opportunity comes from recognizing that we cannot continue along the same path that got us here and there are numerous alternatives to our current way of doing things.

[ECONOMICS AND THE LAWS OF NATURE]
The challenge now is to get things right. We live in a world constrained by fundamental physical and biological laws. We are limited by gravity and the speed of light, which means we cannot float in anti-gravity devices or travel through the universe faster than light. The First and Second Laws of Thermodynamics inform us that while energy exists in different states (heat, light, chemical, electrical), matter and energy can be neither created nor destroyed. As that energy is transformed from one state to another (from the potential energy of a watchspring to the mechanical energy of the spring unwinding, for example), more energy must be added to keep it going. There is no perpetual-motion machine.

As biological creatures, we have absolute requirements from the biosphere. Deprived of air for a few minutes, water for a few days, or food for a few weeks, we perish. Tainted with toxic chemicals, those same life-supporting elements cause us to sicken, even die. Every bit of the energy in our bodies is generated by photosynthesis. These are immutable realities defined by our biological makeup.

Centuries ago, people believed in dragons, demons, and monsters and sacrificed jewels, gold, and even human beings to appease them. Today we know these demons and monsters were figments of our imagination. But now, like the dragons and demons of old, the economy has come to be treated as if it were a real entity before which we must all bow down and sacrifice things of value like the air, forests, oceans, and entire ecosystems.

Capitalism, free enterprise, the economy, markets, corporations, and currency are not natural elements or forces of nature. The last thing we should do is resurrect our superstitious beliefs and, as people did in the past, shovel buckets of money to appease what is our own creation when it shows signs that it isn't working. We created them and if they are not working, we can change them.

Business people and politicians tell me that I "have to be realistic; the economy is the bottom line." I was once told by a Canadian minister of the environment that "environmentalists should understand we can't afford to protect the environment if we don't have a strong growing economy." So even a minister whose primary job is to protect the environment bows before the economy as the highest priority.

Ever since becoming the prime minister of Canada, Stephen Harper, like former U.S. president George W. Bush and former Australian prime minister John Howard, has said that we cannot even try to reduce greenhouse gas emissions to meet Kyoto because of its negative effects on the economy. They seem to forget that economics and ecology are both based on the Greek word *oikos,* meaning household or domain. Ecology is the study of home, while economics is its management. Ecologists try to determine the conditions, laws, and principles that enable life to survive and flourish. By elevating the economy above ecological principles, we seem to assume that we are immune to the laws of nature. But since all of our most fundamental needs for survival come from the biosphere, raising this human-created entity—the economy—above that reality is suicidal. And to make matters worse, the economic construct we have now globalized is fundamentally flawed, so even if it were subsumed by ecology, it would remain a destructive construct.

Economos is the rules and regulations for running the domain while eco-logos is the reason for it all, the underlying principle, the spirit. Normally, the logos should determine the nomos, but in the late 20th century, this is not the case.—SUSAN GEORGE, political economist

In battles over forests, coral reefs, or wetlands, for example, environmentalists are often forced to argue against those who wish to "develop" them from an economic point of view. Thus, when the forest industry claims the economic benefits from logging of jobs, profit, lumber, and pulp, environmentalists must counter with the possible economic value of potential medicines, new genetic material for crops, harvesting of fruit and nuts, and tourism. Environmentalists seldom win in this forum.

Yet the reality is that every ecosystem performs "services" that maintain the conditions necessary for all life. Forests store and pump out vast quantities of water, thereby modulating weather and climate; they remove carbon dioxide from the air and generate oxygen by photosynthesis; they inhibit erosion; and they provide essential habitat to countless other species. Such "ecosystem services" are priceless: they keep the planet healthy. Yet conventional economics ignores these services as "externalities."

It is not Christ who is crucified now, it is the tree itself, and on the bitter gallows of human greed and stupidity. Only suicidal morons, in a world already choking with death, would destroy the best natural air conditioner creation affords... —JOHN FOWLES, writer

For terrestrial animals like us, one of nature's most important services is the pollination of flowering plants. Insects, small mammals like bats and mice, birds, wind, and so on all play a role in how plants exchange genetic material and reproduce new generations of plants. When apiarists began to report the mysterious disappearance of large numbers of honey-bees due to what is now called colony-collapse disorder, I felt a real chill of fear. If major pollinators like bees disappear, whole chunks of Earth's ecosystems will collapse. Yet in our economic system, insecticides can be sprayed over vast areas to control the handful of insects that are pests to us without paying any heed or price for the loss of the many other insect species, including pollinators that keep terrestrial ecosystems healthy.

When a newly elected Ontario government decided that a "greenbelt" of land that encircles the city of Toronto should be protected from exploitation, developers loudly decried the loss of revenues that would result. Development is always a siren's call to local government, with its promise of immediate jobs and revenue. It's the driving force behind the all-out development of Alberta's tar sands, widely decried as the "dirtiest" form of energy. Because nature's services are often ignored or not factored into the equation, development invariably wins the economic argument. However, it is possible to put a price on services that nature performs, such as water filtration, flood control, climate stabilization (by removing and sequestering CO_2), waste treatment, providing habitat for wildlife, and recreation. A study by the David Suzuki Foundation entitled *Ontario's Wealth, Canada's Future* conservatively estimates that $2.7 billion worth of ecosystem services (which economists ignore as irrelevant) was obtained annually in the 1.8-million-hectare greenbelt. That comes to $3,487 per hectare every year. Developers provide a one-time injection of money, and thereafter the property becomes a drain on government services and resources.

The services performed by nature should be our highest concern for our own self-interest because they enable animals like us to survive and flourish, but they are ignored by conventional economists. While we do charge up to $90 per tonne of garbage put in landfills, the notion of putting a price on carbon released in the atmosphere remains anathema in North America. Let's put the *eco* back into economics.

Our lives are absolutely dependent on clean air, clean water, clean soil, clean energy, and biodiversity, and without them, we sicken and die. Yet the economy is built on extracting raw materials from the biosphere and pouring wastes back into it without regard to those services. Disregarding nature and her services is ultimately suicidal, yet it's exactly what conventional economics does. (The tragedy—and the opportunity—is that if done properly, many renewable resources can be harvested indefinitely.)

What if our economy were organized not around the lifeless abstractions of neoclassical economics and accountancy but around the biological realities of nature?
—PAUL HAWKEN, businessman

From a Wartime
Economy to Peace

THE STOCK MARKET crash of 1929 plunged millions around the world into the Great Depression, ruining lives and scarring those who lived through it. What pulled the American economy up was World War II and the enormous productivity of U.S. industry as the country focussed on creating weapons for the Allied cause. But when the war ended, how would a wartime economy function in a world of peace?

A year after the war's end, President Truman established the Council of Economic Advisors, which recommended that consumption become the driver of the economy. In 1953, the chairman of President Eisenhower's Council of Economic Advisors stated that the American economy's "ultimate purpose" was "to produce more consumer goods."

The solution was perhaps best articulated by Victor Lebow, an economist and retailing analyst, in "The Real Meaning of Consumer Demand," published in the Spring 1955 issue of *Journal of Retailing.* To maintain economic growth, Lebow wrote:

Our enormously productive economy demands that we make consumption our way of life, that we convert the buying and use of goods into rituals, that we seek our spiritual satisfactions, our ego satisfactions, in consumption. The measure of social status, of social acceptance, of prestige, is now to be found in our consumptive patterns. The very meaning and significance of our lives today is expressed in consumptive terms. The greater the pressures upon the individual to conform to safe and accepted social standards, the more does he tend to express his aspirations and his individuality in terms of what he wears, drives, eats—his home, his car, his pattern of food serving, his hobbies.

These commodities and services must be offered to the consumer with a special urgency. We require not only "forced draft" consumption, but "expensive" consumption as well. We need things consumed, burned up, worn out, replaced, and discarded at an ever increasing pace. We need to have people eat, drink, dress, ride, live, with ever more complicated and therefore, constantly more expensive consumption. The home power tools and the whole "do-it-yourself" movement are excellent examples of "expensive" consumption.

If a product is built to be durable, eventually the market will be saturated. But if obsolescence, changing fashion, or disposability is designed into those products, an unending market is fuelled by endless demand. By the end of the twentieth century, 70 per cent of the U.S. economy was built on consumption. But all of the raw materials for our products come from the biosphere, and when those products are discarded, they are returned to the planet as waste. And there are unintended consequences.

But there's a second major problem with conventional economics. Economists believe that human inventiveness and productivity are the very core of the economy, and no one has stated this more aggressively than economist Julian Simon: "There is no physical or economic reason why human resourcefulness and enterprise cannot forever continue to respond to impending shortages and existing problems with new expedients that, after an adjustment period, leave us better off than before the problem arose."

And since human imagination and inventiveness are limitless, economists believe there are no barriers to constant expansion of the economy. Thus, economic growth has become the barometer of political and corporate success. Again, listen to Julian Simon:

More people and increased income cause problems in the short run... These problems present opportunity and prompt the search for solutions... In the long run the new developments leave us better off than if the problems had not arisen... We have in our hands now—actually, in our libraries—the technology to feed, clothe, and supply energy to an ever-growing population for the next 7 billion years... Even if no new knowledge were ever gained... we would be able to go on increasing our population forever, while improving our standard of living and our control over our environment.

Others express a different point of view. Physicist Albert Bartlett points out the harsh reality that "in all systems, growth is a short-term, transient phenomenon." He emphasizes that we cannot ignore the mathematical certainty that relentless growth is an impossibility.

Ask any politician or corporate executive how well he or she, the government, or business did last year, and chances are the answer will be based on growth in market share, profit, or GDP.

This obsession for maximizing profits to shareholders has got to be seen as abusive, as dangerous, and as one of the most appalling situations on this planet. Because it makes for criminal behaviour.—ANITA RODDICK, founder of The Body Shop

But by itself, growth is nothing. It merely describes the state of a system. How can growth be the goal or purpose of an economy? It is the context within which growth occurs—what caused the growth, what the increased economy is to be used for, what the impact of that growth will be on people and ecosystems—that is all-important. For example, our bodies require constant production of blood cells to replace the ones that die. But unbridled growth in any part of the body, even of blood cells, is, of course, cancerous and is impossible to sustain in the human body or any system within the biosphere.

And by focussing on growth, we fail to ask the most important questions, like "How much is enough?" "What are the limits?" "Are we happier with all this growth?" and "What is an economy for?"

The late Carl Sagan informed us that if Earth were reduced to the size of a basketball, the biosphere would be thinner than a layer of varnish painted over it. Within that thin layer, all life has flourished, and nothing in it can grow forever.

Let me show you why.

Steady growth over time—whether the amount of garbage produced, the size of a city, or the population of the world—is called "exponential growth," and anything that increases exponentially has a predictable doubling time. Something that grows at 1 per cent a year, for example, will double in seventy years; at 2 per cent, in thirty-five years; at 3 per cent, in twenty-four years; and so on.

Imagine a test tube full of bacterial food. One bacterium is added to the test tube and begins to grow and divide every minute. (The bacterium represents us and the test tube, the planet.) At time zero, there is one cell; at one minute, there are two; two minutes, four; three minutes, eight; and so on. That's exponential growth. At sixty minutes, the test tube is full of bacteria and there is no food left.

When is the test tube half, or 50 per cent, full? At fifty-nine minutes, of course; yet one minute later, the tube will be completely filled. At fifty-eight minutes, it's 25 per cent full; at fifty-seven minutes, it's 12.5 per cent full. At fifty-five minutes, the test tube is only 3 per cent full. If at that moment, one of the bacteria points out they have a population problem, others would jeer, "What have you been smoking? Ninety-seven per cent of the test tube is empty, and we've been around for fifty-five minutes!" Yet they would be five minutes from filling it.

Let's say that at fifty-nine minutes, the bacteria belatedly realize they have only a minute left and pour money into scientific research. But the test tube is all they have. They can no more increase the amount of food and space than we could increase the amount of air, water, soil, or biodiversity on Earth. This is not speculation or hypothesis; it is straight mathematical certainty. Even if, by some miracle, those bacterial scientists were able to create three new test tubes of food to quadruple the amount of food and space (the bacterial equivalent of our discovering three more Earth-like planets), continued exponential growth would fill them all in two minutes after filling the first at the sixtieth minute. So the demand for relentless, uncontrolled growth is a call to race down a suicidal path.

Our home is finite and fixed; it can't grow. And if the economy is a part of and utterly dependent on the biosphere, the attempt to maintain endless growth is a delusional fantasy. Every scientist I've discussed this with agrees with me that we are already past the fifty-ninth minute. A report by the World Wide Fund–UK examined the length of time it takes for nature to replenish renewable resources (trees, fish, soil, etc.) that all humans remove in a year. So long as those resources are restocked in a year or less, that situation should be sustainable indefinitely. The report concluded that it takes 1.3 years to replace what humans exploit in a year, and that deficit has been going on since the 1980s. In other words, rather than living on the biological interest, we are drawing down on our basic natural capital.

When I say we are past the fifty-ninth minute, politicians and business executives become angry. "How dare you say that," they demand, "when our stores are filled with goods and our people are healthier and living longer?" I make no apology for what I say. We have created the illusion that everything is fine by using up what should be the rightful legacy of our children and grandchildren.

I spent the first six years of my life in Vancouver, and I vividly remember Dad rowing around Stanley Park to catch sea-run cutthroat trout and jigging for halibut off Spanish Banks. Today when my grandchildren call and beg me to take them fishing, I can't take them where I went as a child because no fish are left there.

Carp caught in the Thames River in
London, Ontario, late 1940s

The GDP:
A Flawed Instrument

"Much of what we now consider economic growth, as measured by GDP, is really the fixing of blunders from the past and the borrowing of resources from the future."—CLIFFORD COBB, theologian, and TED HALSTED, founder, New America Foundation

AN ECONOMY MUST exist to improve the quality of life for people. When growth becomes the primary goal of an economy, the instruments developed to measure economic success are based on measuring growth. But the leading measurement today, the GDP (gross domestic product), is merely a sum of national spending, with no distinctions between transactions that add to well-being and those that diminish it. The result, as crusader Ralph Nader says, is that every time there is a car accident and someone is killed or badly injured, the GDP goes up because you need ambulances, doctors, caskets, lawyers, and so on. If you live in a neighbourhood where crime is increasing so that you have to buy more insurance, locks for doors and windows, burglar alarms, weapons of self-defence, and so on, all of these items add to the GDP, but clearly the quality of life is not improving.

In 1995, the organization Redefining Progress created an instrument called the Genuine Progress Indicator (GPI) as an alternative to the GDP. The GPI considers income distribution, adding money spent if the poor receive a greater share of the economy and subtracting when the rich get more. The value of household work, volunteering, and higher education, which do not normally contribute to GDP, is added to the GPI. The cost of crime, resource depletion, pollution, environmental damage, and disposable goods is subtracted from the GPI.

A plot of the GDP and the GPI from 1950 to 2004 shows the GDP progressing relentlessly upward, with occasional dips here and there to signal recession, while the GPI rises more slowly until it peaks around 1970 and from then begins to drop. The GPI corroborates what we know intuitively, that more and more of us are working longer and harder without improving the quality of our lives or finding more leisure time for our children, hobbies, or community.

"The [GDP] counts air pollution and cigarette advertising, and ambulances to clear our highways of carnage… Yet [it] does not allow for the health of our children, the quality of their education, or the joy of their play… It measures neither our wit nor our courage; neither our wisdom nor our learning; neither our compassion nor our devotion to our country; it measures everything, in short, except that which makes life worthwhile."
—ROBERT F. KENNEDY, politician

All over the world—in the Amazon, Australia, the Serengeti, the Arctic—I have sought out elders and asked, "What was it like here when you were a child?" Everywhere the answer has been chillingly similar: "It used to be so different." "There used to be trees as far as you could see." "The rivers used to teem with fish." "The skies once were blackened by birds at certain times of the year."

Once it was said as development cleared forests, fish, or birds, "There's plenty more where that came from." Around the globe, elders today are living records of enormous changes that have occurred in the span of a single human life, attesting to the fact that there isn't plenty more. People in industrialized nations often flippantly dismiss concerns about social and environmental degradation caused by development by saying, "That's the price of progress." But is it progress to create technologies for human use that undercut the conditions within the biosphere that keep us alive? Is it progress to use up what should be the rightful legacy of our children or to leave them to deal with problems that we have created?

We must convince each generation that they are transient passengers on this planet earth. It does not belong to them. They are not free to doom generations yet unborn. They are not at liberty to erase humanity's past nor dim its future.—BERNARD LOWN, co-founder, International Physicians for the Prevention of Nuclear War

It used to be understood that we have a sacred duty to pass on to future generations a world that is as rich as or richer than the one we came into. We are no longer living up to that obligation. Traditionally, when North American Aboriginals had to make important decisions, they first reflected back on their ancestors and then thought ahead to how their decision might affect the seventh generation after them. Today we know our actions are drawing down on the resources and conditions of future generations, and we need that perspective from the past and into the future.

For most of our existence, people knew that we were deeply embedded in nature and that our very survival depended on nature's generosity.

We understood that everything in the world was connected, that what we did had repercussions, and that therefore every act was laden with responsibility.

Nature was our touchstone and our reference point and dictated the way we interacted with it. But as economics and politics have increasingly come to dominate our decisions and actions, we have lost our sense of place in the world and our reverence for nature. We need a new relationship with the planet that is, in fact, our ancient understanding.

[THE LIMITS OF REDUCTIONISM]
Science itself must also take some of the blame for the
loss of reverence and respect for nature. Like most scientists I know, I was drawn into science by the allure
of nature. I became bonded to the natural world during World War II, when Japanese Canadians were
incarcerated in camps in the Rocky Mountains, in a
spectacular area that is now Valhalla Provincial Park.
There I fished, gathered mushrooms and flowers for my
mother, and encountered without fear wolves, bears,
and elk.

*Those who contemplate the beauty of the Earth find
reserves of strength that will endure as long as life lasts.
There is symbolic as well as actual beauty in the migration of birds, the ebb and flow of tides, the folded bud
ready for spring. There is something infinitely healing
in the repeated refrains of nature—the assurance that
dawn comes after the night and spring after the winter.*
—RACHEL CARSON, biologist

AFTER THE WAR, when my parents became farm
workers in southern Ontario, I fell in love with insects,
particularly beetles, and spent countless hours wading
through my magical swamp. When I became a geneticist, it was to study heredity in an insect, the fruit fly.
Astronomer Carl Sagan told me that as a child he was
drawn to the mysteries of the heavens. Paul Ehrlich

*Trout from Beatrice Lake, now part
of Valhalla Provincial Park, 1943 or 1944*

was attracted to biology by his enchantment with butterflies, and Ed Wilson came to ecology because of his
early fascination with reptiles. So it seems strange that
while mystery, wonder, and awe imbued our view of
the world and drew so many of us into science, as we
write up our reports, we expunge all trace of passion
and emotion in the name of objectivity. Yet it is more
important than ever to maintain that sense of wonder,
awe, and reverence as we approach the natural world
to seek ways to repair the disruptions within the biosphere that we have caused and that make the world a
more difficult place for our own survival.

As scientists, many of us have had profound experiences of awe and reverence before the universe. We understand that what is regarded as sacred is more likely to be treated with care and respect. Our planetary home should be so regarded. Efforts to safeguard and cherish the environment need to be infused with a vision of the sacred.—National Religious Partnership for the Environment

But science, that wondrous achievement of the human brain, obliterates wonder and awe, the sense of the sacred or the profane, when it focuses on parts of nature—a powerful methodology called reductionism. This approach assumes that the cosmos works like an immense machine, a "clockwork mechanism" whose secrets can be revealed by examining the parts and then piecing them back together.

But in focusing on the parts, we lose all sense of the whole, and today we know that the whole is greater than the sum of its parts. That's because when combined, the pieces interact, and properties emerge from their interaction that cannot be anticipated from the characteristics of the individual parts. So, for example, the properties of atomic hydrogen and atomic oxygen cannot be used to anticipate the "emergent properties" exhibited by their combination in a molecule of water.

I realized the limitations of reductionism in 1962 when I read Rachel Carson's seminal book *Silent Spring*, which examined the ecological ramifications of DDT and other pesticides and galvanized the global environmental movement. Her book taught me that in focusing on parts of nature, in examining them in controlled conditions in flasks and growth chambers, we study artifacts, grotesque simplifications of the real world, scrubbed of the context of weather, climate, and seasons, devoid of variations in temperature, humidity, and light.

Until I read *Silent Spring*, I had assumed that my experiments were miniature replicas of the real world. Now I understood that while studying bits of nature under controlled conditions can provide powerful insights, we had to be very cautious in extrapolating from those tests to the real world.

Galvanized by *Silent Spring*, we realized the consequences of powerful technologies and exploding demands on nature: disappearing forests, pollution, threatened species. Environmentalists were inspired to try to protect whales, seals, grizzly bears, and other threatened animals and to fight to stop pollution and clear-cut logging. But these were immediate problems that demanded the immediate attention of the young movement, and environmentalists were dragged into dealing with these symptoms rather than the root causes of our destructiveness.

Losing the Past
by Shifting Baselines

HUMAN BEINGS ARE not bound to an ecosystem or territory by instinct or physiology. Our large and complex brain
enabled us to learn about and exploit our surroundings and
to shape our environment so that we could live within it. Our
adaptability has enabled us to survive and flourish in environments as extreme as deserts and the Arctic tundra, as
well as prairies, wetlands, forests, and mountain slopes.

Today that adaptability has reached astonishing levels as
technological development and change accelerate. Today's
latest cellphone, digital camera, or iPod, a miracle of technological ingenuity, becomes tomorrow's soon-to-be-forgotten
discard. But there is a huge cost to our short memories.
Throughout time, elders have been a reminder of what
once was, and their stories set the bar on abundance of fish,
turtles, trees, or birds.

I live on a street that once had a large empty lot with
a small grove of trees on it, where neighbourhood kids,
including mine, spent loads of time playing. But one day the
trees were all cut down, and within months, an apartment
building had replaced the trees. A year later, none of the
occupants of the apartments knew anything about the trees
that had once been there, and the rest of us on the block

soon forgot them too. Without memory, where do we set the baselines as we try to restore nature?

We tend to think that what we have experienced is more or less the way it has always been. A documentary on fishing, *Empty Oceans, Empty Nets,* shown on PBS in 2002, featured an interview with a young skipper on a swordfish boat from Boston who stated that there are still plenty of swordfish. Based in Boston, she travels up to Newfoundland, where she reported hearing that a 200-pound swordfish had been caught. "There are still big ones," she said. The film then cut to an interview with a grizzled fisherman who must have been in his eighties. He recounted that he used to fish just 5 or 6 miles out of Boston and would throw back anything under 200 pounds! Two fishers with radically different baselines. To the young skipper, a trip all the way to Newfoundland was standard procedure, while a 200-pounder was a big fish. (In fact, the average size of swordfish before 1963 was 266 pounds; it had fallen to 133 pounds in 1973, and to 90 pounds in 1996.)

When *The Nature of Things* broadcast a program on grizzly bears, I was astonished to learn that they are not the mountain creatures we think they are today. Grizzlies were once plains animals that preyed on the vast herds of bison that moved along the central corridor of North America. Grizzlies ranged from the Pacific coast to Ontario, from the Arctic down to Texas. But as bison and the grizzlies were deliberately extirpated, their range shrank to the remote areas of mountains. And there are many more examples like these.

In British Columbia, for example, one group might be focused on declining salmon populations, another on protection of endangered species—wolves, bears, and eagles—and still others on clear-cut logging. But the forest, salmon, eagles, wolves, and bears are exquisitely interconnected in a web of mutual dependence.

Those salmon go all the way out into the ocean. And this is what the old people would think: these salmon, their bodies are sacred. They're gathering all these foods, just as we do. And they're bringing them back to nourish us. But the food also nourishes the bear; it nourishes the eagle, the cougar, the animals, the bugs, the nutrients, the microfauna, and that's what the next generation of salmon are going to live off.—DON SAMPSON, Oregon First Nations leader

The thin strip of land pinched between the Pacific Ocean and the coastal mountains and extending from Alaska south to California supports a temperate rainforest that has some of the largest trees on Earth. But how can such large trees grow when the heavy rains that create the rainforest wash away soil nutrients such as nitrogen, which is a vital component in the soil for the growth of trees? The answer reveals the exquisite interconnectedness of every part of nature.

Up and down the Pacific coast, thousands of rivers and streams have supported salmon populations that are born in fresh water, go out to the ocean for years, and return to spawn and die in their natal waters. It has long been known that the salmon need the forest, because when watersheds supporting the fish are clear-cut, salmon populations plummet and disappear. Salmon are very sensitive to temperature and need the forest canopy to shade the water to keep the temperature down. The tree roots cling to soil to prevent it from eroding into spawning gravels, and the newly hatched salmon are fed by the forest as they make their way to sea.

Through his work tracing nitrogen from the ocean in the soil of the rainforest, University of Victoria biologist Tom Reimchen has shown that salmon are the largest source of nitrogen fertilizer delivered to the forest. After two to five years in the ocean, depending on the species, salmon return to their birthplace rich in marine nitrogen, which can be distinguished from terrestrial nitrogen. Eagles, wolves, and bears consume the salmon and then spread the marine nitrogen when they urinate and defecate throughout the forest.

Reimchen showed that a bear will take over six hundred salmon in a season, hiking up to 150 metres away from the river's edge to eat in privacy. It eats about half of the fish, leaving the rest of the carcass and going back to the river for another. The remains are eaten by ravens, salamanders, insects, and other creatures. Most of the salmon carcass is consumed by fly maggots, which grow to full larval size before dropping into the forest litter to overwinter. The following spring the marine-nitrogen-laden adult flies emerge from the pupal condition just in time to feed birds passing through from South America on their way to their Arctic nesting grounds.

It turns out that bears are a major vector for spreading marine nitrogen through a host of other animals and plants. Many of the spawned-out salmon die and sink to the bottom of rivers, where they are soon covered in a thick coat of fungi and bacteria, which in turn are consumed by insects and other invertebrates. When baby salmon emerge from the spawning gravel, the waters are filled with marine-nitrogen-rich food; so in dying, the parents prepare a feast for their offspring. Salmon, forest, birds and bears, ocean, air and land, northern and southern hemispheres are all interconnected through a web of interdependence.

But humans view the situation very differently. The trees fall under the "management" of the minister of forests; the salmon under the ministers of tourism (sport fishers), Indian and northern affairs (native food fishers), and fisheries and oceans (commercial fishers); eagles, bears, and wolves under the minister of the environment; the water under the minister of agriculture (for irrigation) or energy (hydro power); the rocks and mountains under the minister of mining. Thus, we fragment what is a single interconnected system into separate components, pulling at individual strands of a vast fabric of interdependence and thereby ensuring that we will never manage things sustainably. We must learn to look at the big picture and to see the interconnectedness of all things.

[WORKING TOGETHER]

We must also join together to deal with the crisis we are facing. In the 1950s, when I was beginning my senior year at Amherst College in Massachusetts, I was witness to a remarkable response to a crisis.

On October 4, 1957, the Soviet Union electrified the world by announcing that it had launched *Sputnik*, a basketball-sized satellite, into orbit around the planet. The Cold War between the superpowers was in full force. As *Sputnik*'s electronic beep taunted the West, the United States air force, navy, and army tried to send a grapefruit-sized satellite into space aboard their different rockets, only to fail miserably as each one blew up before or shortly after launch.

While the U.S. struggled to catch up, Russia announced one space first after another—the first animal (a dog, Laika), the first man (Yuri Gagarin), the first team of cosmonauts, the first woman (Valentina Tereshkova) to orbit the Earth. The USSR was a formidable force. The U.S. began a massive effort to catch

Amherst College graduate,
soon after the launch of Sputnik

up, funnelling billions of dollars into government labs, universities, and students—even foreigners like me.

In 1961, when U.S. president John F. Kennedy announced plans for Americans to be the first to land on the moon, no one denounced it as a preposterous dream, complained that the Russians were too far ahead, or accused Democrats of being fiscally irresponsible. Instead, the American government and citizens joined together to meet the Soviet challenge. They mobilized an all-out effort and became the first and only nation to put astronauts on the moon.

Decades later, the U.S. had reaped enormous unexpected benefits. That research and effort led to cellphones, stationary satellites, around-the-clock news channels, GPS, and a virtual stranglehold on Nobel prizes in science. All this happened because the U.S. faced the daunting challenge of Soviet supremacy head on and made a commitment to pour everything into catching up and passing them.

Until one is committed, there is hesitancy, the chance to draw back... The moment one definitely commits, then Providence comes too. All sorts of things occur to help one that would never otherwise have occurred... Whatever you can do or dream, you can begin it. Boldness has genius, power and magic in it. Begin it now.—JOHANN WOLFGANG VON GOETHE, writer

That is what is needed now to confront our enormous ecological challenges—a joining together in a common goal and a commitment to meet that goal. And if we do those things, we can be assured that there will be huge unexpected benefits.

A Vision for

THE FUTURE

B Y THE late 1970s, I could see that we needed a different perspective on environmental problems, and for me, that came when I received a letter from Jim Fulton, the member of Parliament representing Skeena, a huge riding in the northwest corner of British Columbia. It began, "Soozook, you should do a film about the fight over logging in Haida Gwaii."

I quickly learned that for years, in the chain of islands extending south from the Alaska Panhandle, a battle had raged pitting environmentalists and Haida, the inhabitants of the islands, against politicians and forest companies. In the early 1980s, the Haida, led by president Miles Richardson, decided to draw the line and stop the logging trucks from entering Windy Bay, a pristine watershed of some six thousand acres that was sacred to the Haida.

The land we hold in trust is our wealth. It is the only wealth we could possibly pass on to our children... Without our homelands, we become true paupers.
—ANNETTE HELMER, Inuit elder

I flew to Haida Gwaii and interviewed forest company executives, politicians, environmentalists, loggers, and Haida. I met Guujaaw, a young Haida artist and carver who had led the struggle against logging for years. "What would happen if the trees were all cut down?" I asked. His reply sailed right over my head: "Then we'll be like everyone else, I guess." Days later, as I watched the rushes in Vancouver, I suddenly saw that his simple statement opened a window on a radically different way of seeing the world.

Guujaaw and the Haida do not see themselves as ending at their skin or fingertips. Of course they would still be around physically if the trees were all gone, but a part of what it is to be Haida would be lost. The trees, fish, birds, air, water, and rocks are all a part of who the Haida are. The land and everything on it embody their history, their culture, the very reasons why Haida are on this earth. Sever that connection and they become "like everybody else."

Since my first encounter with Guujaaw, I have met Aboriginal people in other parts of the world and witnessed that same attachment to place. And I have become a student of their perspective on our relationship with the planet.

Whether it's in the Amazon, the Serengeti, or the Australian outback, Aboriginal people speak of Earth as their mother and tell us we are created by the four sacred elements: Earth, Air, Fire, and Water. I realized that we had defined the problem incorrectly. I had pressed for laws and institutions to regulate our interaction with the environment when, in fact, there is no environment "out there," separate from us; I came to realize that we *are* the environment.

Leading science corroborates this ancient understanding that whatever we do to the environment or to anything else, we do directly to ourselves. The "environmental" crisis is a "human" crisis; we are at the centre of it as both the cause and the victims.

[A CHANGE IN PERSPECTIVE]

Let's look at the world through different eyes. What do we really require to live full, rich, healthy lives? Rather than being separate or apart from the rest of nature, we are deeply embedded in and utterly dependent on the generosity of the biosphere.

Biophilia

LOVE IS A powerful force that motivates behaviour in social animals and helps provide an evolutionary advantage for their offspring. But as any owner of a dog, cat, bird, or horse knows, it would be difficult not to call what we feel for these pets love. Many avid gardeners assure me their obsession is driven by their love of plants or love of working with soil. Are we simply confusing language, or do we really mean we "love" these other species?

The ecologist Edward O. Wilson tells us we know that human beings evolved out of a state of nature and have lived in the company of many other species on whom we depend for our survival as food and companionship. So he has coined the word *biophilia* (*bio*—life and *philia*—love) to signify what he believes is an innate need to "affiliate with other species." In other words, our need to be with other species is built into our very genome. Numerous studies corroborate this notion. People in hospitals, old age homes, and similar institutions respond positively when animals and plants are brought into their rooms. Cancer patients feel better when they are out in nature. Prices of homes indicate our preference to be near water and forests. Stress is alleviated by walks in woods or parks. And so on.

As more and more of humanity moves to big cities, we lose that direct contact with nature that our ancestors always had. Often we demonize nature, saying to our children, "Don't touch" when they confront an insect or frog, or "Don't bring that in here" when they have insects, some soil or pond water in a container.

When I was a hormone-ravaged teenager, my sanctuary from awkward social interactions was an enchanted place, a swamp only a few miles by bike from our house in London, Ontario. I remember so many magical hours spent wading through the waters with a net to collect frog's eggs, insects, and any other treasures I might come across. It was in that swamp that I first caught a soft-shelled turtle, spotted a bittern with its beak aiming straight up to blend in with the reeds, and watched a raccoon digging clams. Today that swamp is entombed by a massive parking lot that serves a large shopping centre. Where do children today find their outlet for their biophilic needs?

[AIR]

The first thing every infant needs at birth is a breath of air to inflate its lungs and then announce its arrival to the world. From that moment on to his or her dying breath, every person needs air fifteen to forty times a minute.

Lungs are made up of some 300 million alveoli, capsules lined by a three-layered membrane called surfactant that reduces surface tension so that the air sticks to it. Oxygen (and whatever else is in that breath) enters the bloodstream, where hemoglobin molecules in red blood cells pick it up and with each beat of the heart deliver it throughout the body. Even when you breathe out, about half the air remains in the lungs; otherwise they would collapse. In other words, we cannot draw a line that delineates where air ends and we begin because air is in us, fused to our lungs and circulating in our bloodstream. We are air.

When you exhale, the breath that leaves your nose quickly mixes with the air and goes straight up the nose of anyone nearby. If I am air and you are air, then I am you. And we are embedded in the matrix of air not just with all other people on the planet but also with the trees and birds and spiders and snakes.

The American astronomer Harlow Shapley once wondered what happens to a breath of air. Ninety-eight per cent of air is oxygen and nitrogen, which we need for our bodies' metabolic processes, so when we exhale, varying amounts remain in us. But 1 per cent of air is argon, an inert gas, which enters our bodies when we inhale but then comes back out when we exhale. Thus, argon is a good "marker" for a breath of air.

Shapley calculates that there are 3×10^{19} argon atoms in a breath of air. That's three followed by nineteen zeros! Within minutes, a breath has mixed with the surrounding matrix of air, and eventually it is blown all over the planet. A year later, according to Shapley, wherever we may be, every breath will carry about fifteen argon atoms that came from a breath a year earlier! So every breath we take contains argon atoms that were once in the bodies of Joan of Arc and Jesus Christ; every breath contains argon atoms that were once in dinosaurs 65 million years ago; and every breath will suffuse all life far into the future.

Our next breath, yours and mine, will sample the snorts, bellows, shrieks, cheers, and spoken prayers of the prehistoric and historic past.—HARLOW SHAPLEY, astronomer

Air is more than just a physical component of Earth; it is a *sacred* element giving life to all terrestrial organisms, linking all life in a single matrix, and joining past, present, and future in a single flowing entity.

Our great boast is the possession of intelligence, but what intelligent creature, knowing the critical role of air for all life on Earth, would then proceed to deliberately pour toxic materials into it? We *are* air, so whatever we do to air, we do to ourselves. And this is true of the other sacred elements.

[WATER]

A visitor from another galaxy would surely call our planet Water, not Earth. Seventy-one per cent of the planet's surface is covered by oceans. If the globe were a perfectly smooth sphere, water would cover it to a depth of 2.7 kilometres. The air is filled with water vapour that condenses as clouds. Above the great Amazon rainforest, trees pull water from the ground and transpire it upward, where it flows in great rivers of vapour toward the Andes.

Every person in the world is at least 60 per cent water by weight. We are basically blobs of water with enough organic thickener mixed in to prevent us from dribbling away on the floor. The hydrologic cycle of evaporation, condensation, and rain ensures that water cartwheels around the planet. We are part of the hydrologic process. Every drink we take has water molecules that evaporated from the canopies of every forest in the world, from all of the oceans and plains.

Again, we say we are intelligent, but what intelligent creature, knowing that water is a sacred, life-giving element, would use water as a toxic dump? We are water, and whatever we do to water, we do to ourselves.

[EARTH]

Soil is alive, created by life and supporting an astounding array of organisms. Microscopic examination of soil reveals a world where hard and soft, liquid and gaseous combine in an alchemy of life where organic and inorganic, animal, vegetable, and mineral all interact. Death turns into life, which in turn feeds more life; every cubic centimetre of soil teems with billions of micro-organisms. Soil varies enormously around the world in both biological and inorganic makeup, but in general about 45 per cent of soil is minerals, 25 per cent is water, 25 per cent is air, and only about 5 per cent by volume is organic matter, living and dead.

Nitrogen, water, and carbon flow through the community of organisms in the top 5 to 10 centimetres of soil. That community ranges from the microscopic to those organisms visible to the human eye. A single gram of soil from the root zone of plants may contain a billion bacteria, comprising four to five thousand species, although some experts suggest it may be more like forty thousand species. We really have no idea how many soil species there might be, but guesses are 2 to 3 million species of bacteria and 1.5 million fungi, of which a mere 2 to 5 per cent have been described and named. A square metre of soil may support 10 million nematodes, a billion protozoa, and 200,000 to 400,000 springtails and mites. Then there are the bigger creatures, like earthworms, ants, termites, millipedes, woodlice, beetles, and insect larvae.

Every bit of the food we eat for our nutrition was once alive, and most of it comes from the soil. We take the carcasses of plants and animals, tear them apart, and incorporate them into our very being. We are earth.

We say we are intelligent, but what intelligent creature, knowing the role the earth plays in constructing our very bodies, would then proceed to use the earth as a dump for our waste and toxic material?

The soil is virtually a living organism. It's not just a bunch of grains with bugs walking through them. It is a mass of organic, living material in an organic matrix. It's dynamic. It's full of life. And it does not produce anything for human beings unless it's sustained in that living condition.—EDWARD O. WILSON, ecologist

We are the earth, and whatever we do to the earth, we do to ourselves.

[FIRE]

Like all other animals, we need energy to move, grow, and reproduce, and every bit of that energy in our bodies is sunlight, transformed by photosynthesis in plants into chemical energy, which can be stored and used when needed. We acquire that energy by eating plants or animals that eat plants and then storing that latent energy in our bodies. And all of the fuel (except nuclear) we burn, from fossil fuels to peat, wood, and dung, is created by that photosynthetic activity. When we need that energy, we burn molecules to liberate the power of the sun for use in our bodies. (My mother used to call me Sonny, but we are all sunny.)

[BIODIVERSITY]

Finally, it is the web of living things, which scientists call biodiversity, that generates the most fundamental needs of all animals—an oxygen-rich atmosphere; water filtered of toxic material; the plants, animals, and micro-organisms that we eat as well as the soil that sustains them; and all fuel we burn as well as oxygen itself, which is needed to create fire. So biodiversity, the web of life itself, should be considered a sacred element, in addition to the other four elements.

The Human Genome Project—the great technological achievement that deciphered the sequence of the

3 billion letters in human DNA—revealed the most astonishing fact: 99 per cent of our genes are identical to the genes of the great apes. They are our nearest relatives.

Thousands of genes in our cells are identical to those in our pet dogs and cats and in birds, fish, insects, and trees. All of life on Earth is our kin. And in an act of generosity, our relatives create the four sacred elements for us.

[LOVE]

Being a human being—in the sense of being born to the human species—must be defined also in terms of becoming a human being... A baby is only potentially a human being, and must grow into humanness in the society and the culture, the family.—ABRAHAM MASLOW, psychologist

We are also social animals, and our most fundamental social need is love. It begins with the flow of hormones between mother and fetus that continues after birth as a mother nurses her baby. In infancy, babies require that love during developmental windows. We need love to learn how to love, to empathize with our fellow beings, to be welcomed into the community of our kind. We must have love to be fully human and to realize our full potential.

If deprived of love, children growing up under conditions of famine, terror, genocide, or war are fundamentally crippled physically and psychically. But a disruption of love can seriously compromise childrens' well-being; they die sooner than normal.

The first act of a newborn child—drinking from the mother's breast—is co-operation, and is as pleasant for the mother as for the child... We probably owe to the sense of maternal contact the largest part of human social feeling, and along with it the essential continuance of human civilization.—ALFRED ADLER, psychiatrist

For decades, scientists have debated whether human behaviour and intelligence are primarily a reflection of genes (nature) or the environment (nurture). In retrospect, it is a ridiculous issue. Genes and their expression cannot occur without an environmental matrix, while an environment without genes is meaningless for living organisms. But now we know that although the primary information in the sequence of letters in a genome remains constant, the expression of those genes is influenced by environmental factors.

One of my scientific contributions was to demonstrate that mutations in DNA can be induced and recovered that produce mutants that are normal at one temperature but are extremely abnormal at a different temperature. So the same gene can have very different expressions, depending on the surrounding temperature. Molecular analysis recently revealed that gene expression can be influenced by modifying DNA without changing the primary sequence of letters within a gene. Stresses such as poverty, malnutrition, and bullying not only induce cellular and structural changes in the brain but cause "changes in cellular and neuronal connections, and most recently, down into lasting changes in how DNA is expressed," according to molecular biologist Thomas Boyce, speaking at an American Association for the Advancement of Science meeting in San Diego in February 2010.

Nicolai Ceauşescu, former dictator of Romania, banned abortion and birth control in an effort to stimulate population growth. After he was executed in 1989, the dimensions of his dictates were revealed as tens of thousands of "orphans," many simply abandoned by their parents, were found in orphanages, many of which were Dickensian in their horrific conditions. Like Harry Harlow's famous experiments on the effects of parental deprivation on baby monkeys, these children in institutions provided critical information about the terrible consequences of the deprivation of love. The children were found to be in the bottom percentiles of physical growth, and many were "grossly delayed" in motor and mental development. They were reported to rock and grasp themselves the way

Harlow's baby monkeys did and grow up with aber-rant social values and behaviour. Hundreds adopted in North America continued to show deficits in size, motor skills, and behaviour. We need love to realize our full human potential.

[SPIRIT]

We are spiritual beings, and we need spirit more than ever. We need to understand that nature gave us birth and is our home and source of well-being, and that when we die, we will return to it.

We needn't be saddled with the impossible weight of managing the entire biosphere, but we must meet the challenge of living in balance with the sacred ele-ments. We are part of a community of beings that are related to us.

We should know that there are forces impinging on us that we will never understand or control. We need sacred places where we go with veneration rather than to seek resources or opportunity.

These, then, are our most fundamental biological, social, and spiritual needs.

Once we understand that those basic needs must be the very foundation of our values and the way we live, that they must be protected for our health and well-being, we can begin to imagine a new way of living in harmony and balance with them.

Learning Sustainability
from Traditional People

AS ECOLOGICAL DEGRADATION accelerates under the pressure of modern technology and economics, some have belatedly recognized the value of indigenous knowledge and perspectives for industrial society. As the groundbreaking UN document *Our Common Future* (1987) stated in reference to Aboriginal people:

Their very survival has depended upon their ecological awareness and adaptation… These communities are the repositories of vast accumulations of traditional knowledge and experience that links humanity with its ancient origins. Their disappearance is a loss for the larger society, which could learn a great deal from their traditional skills in sustainably managing very complex ecological systems. It is a terrible irony that as formal development reaches more deeply into rain forests, deserts and other isolated environments, it tends to destroy the only cultures that have proved able to thrive in these environments.

There is a remarkable congruence of scientific insights and traditional knowledge, as we are beginning to recognize that what were once pristine forests and shores were in fact moulded by sophisticated human activity. Thus, around the world, what were thought to be primary tropical forests have been found to be created by generations of people (some think primarily women) who gathered useful seeds and plants and scattered them around villages while also protecting large areas as nurseries of useful plants and animals. This is called agroforestry.

On the west coast of Canada, scientists have only recently discovered that hundreds of the most fertile areas for clams were created by people as clam gardens hundreds of years ago. And as resources collapse under assault by modern industrial forces, we are turning to those reservoirs of traditional knowledge to gain insights into better management processes.

[REIMAGINING OUR WORLD]

The biggest challenge humanity faces in carving out a better future is to reimagine how we perceive the world, our place within it, and our highest priorities. By creating a vision of what must be, we then determine the way we act.

There is no silver bullet to solve our problems. The change begins with each of us, then with our families, our communities, our country, and the world.

It's all a question of story. We are in trouble just now because we do not have a good story. We are in between stories. The old story, the account of how we fit into it, is no longer effective. Yet we have not learned the new story. —THOMAS BERRY, philosopher

Perhaps we don't need to find a new story but to rediscover an old one. Linguist Nicholas Faraclas presents a different vision of a society:

Imagine a society where there is no hunger, homelessness or unemployment, and where in times of need, individuals can rest assured that their community will make available to them every resource at its disposal. Imagine a society where decision makers rule only when the need arises, and then only by consultation, consensus and the consent of the community. Imagine a society where women have control over their means of production and reproduction, where housework is minimal and childcare is available 24 hours a day on demand. Imagine a society where there is little or no crime and where community conflicts are settled by sophisticated resolution procedures based on compensation to aggrieved parties for damages, with no recourse to concepts of guilt or punishment. Imagine a society... to which the mere fact that a person exists is cause for celebration and a deep sense of responsibility to maintain and share that experience.

I met Faraclas in Port Moresby in Papua New Guinea, where he had lived and taught for many years. In his experience, the society he described above is not a dream or fantasy but is very real.

When the first colonizers came to the island of New Guinea, they did not find one society that exactly fit the above description. Instead, they found over one thousand distinct language groups and many more distinct societies, the majority of which approximated closely the above description, but each in its own particular way. These were not perfect societies. They had many problems. But after some one hundred years of "Northern development"... nearly all of the real developmental gains achieved over the past 40,000 years by the indigenous peoples of the island have been seriously eroded, while almost all of the original problems have gotten worse and have been added to a rapidly growing list of new imported problems.

Our great evolutionary advantage was the ability to lift our sights and look ahead, to imagine the world as it could be and then make the best choices to move toward that vision.

Let's raise our eyes beyond the conventional horizon of a year and ask, "What kind of world would we like to have in a generation?"

How about one in which the air is clean and children no longer have epidemic levels of asthma? I can imagine a world that is covered in forests that can be logged forever because it is being done properly according to principles of ecosystem-based management in which nature and ecology set the rules. When I was a child, we drank the water from any river or lake—how about aiming for that? We used to catch fish and eat them without ever worrying about what toxic chemicals might be in them. That is an achievable goal.

I can imagine a future in which cities are exquisitely adapted to climate, the surrounding landscape and wildlife, and the natural rhythms of the seasons, in which every building captures all of the sunlight and water falling from the heavens, where food is grown on rooftops, where roads are permeable and allow water to percolate back into the earth instead of running through gutters and sewers, where a yard becomes a natural landscape and not a monoculture of grass, and where butterflies flit through gardens in every schoolyard. I can picture a city where cars are rarely needed because all of the action and fun are going on in the streets of the neighbourhoods where we live, work, and play.

When we look ahead and dream a future we'd all prefer, everyone I've discussed it with is in agreement. "Of course," they say, "that would be wonderful. No one would be against that." So we can begin together, and now that we have a target, we know where we want to go. If we set concrete goals to achieve year by year, I believe we can reach what we call "sustainability within a generation."

Economists tell us that we can't realign our economic system to incorporate the kinds of values that people like me hold, that "it's not realistic" to look to a radically different future, that the economy is the bottom line to which everyone and everything must capitulate.

Let's consider this. I live in an oceanfront house in Vancouver. I once received a form letter from a real estate agent announcing that "now is a good time to sell your property and buy up." I began to think about what I would put down as the most valuable parts of the property. Well, my wife's mother and father have lived with us for thirty years, so my children have had Grandma and Granddad right upstairs their entire lives. That's what I would list as a huge value.

My father was a cabinetmaker, and when Tara and I were married, he built a set of cupboards for our first apartment. When we bought our current house, I tore out one of those cupboards and installed it in our kitchen. It didn't fit well, but Dad is with us every time we use it. That went down on my list of values.

My father-in-law is an avid gardener and knowing that I love asparagus and raspberries, he planted them just for me. I put them down on my list.

My best friend, Jim Murray, came out to visit one year and carved a handle for the gate of a fence I was building. Every time I use that gate, I think of Jim. I wrote that down.

My children have brought dead birds, snakes, and squirrels home to bury under the dogwood tree—an animal cemetery—and I wrote that down.

I built a tree house in that dogwood tree, and my children spent many happy hours playing in it. That went down as a value.

Looking over that list, I know those are the things that for me transform a "property" into a "home," and they are priceless. But on the market, they are worth- less. And that's our problem: when we measure every- thing according to its economic value, those things that matter most to us are worthless.

The biosphere is our home. All other species in Cre- ation are not resources, opportunities, or commodities; they are our relatives, and in an act of generosity, they provide our most fundamental needs while also giving us companionship and enriching our lives with beauty, mystery, and awe.

We have to see the world through new eyes, because how we view the world affects the way we treat it.

Ethnobotanist Wade Davis took me to an Andean village in Peru where the inhabitants are taught that the mountain towering over them is an *apu*, or god, that determines their fate. "Contrast the way they treat their sacred mountain with a kid growing up in B.C. who is taught that a mountain may be full of valuable minerals," Wade said.

Is a river the veins of the land or simply potential energy and irrigation? Is a forest a sacred grove or just timber and pulp? Is soil alive or merely dirt? Is a house a home or just a piece of real estate? Is the planet a sacred creation or a world of opportunity? Our values and beliefs shape the way we treat our surroundings.

My father was my great hero and mentor, and in 1994, when he was eighty-five, he learned he was dying of cancer. He wasn't in pain and he was completely lucid. I moved in to care for him during the last month of his life, and it was a wonderful time. He wasn't afraid of death, and with his help, I wrote his obituary:

Carr Kaoru Suzuki died peacefully on May 8. He was eighty-five. His ashes will be spread on the winds of Quadra Island. He found great strength in the Japanese tradition of nature-worship. Shortly before he died, he said: "I will return to nature where I came from. I will be part of the fish, the trees, the birds—that's my reincarnation. I have had a rich and full life and have no regrets. I will live on in your memories of me and through my grandchildren."

He told me when I hear the sigh of wind through the trees or see the flash of a salmon in the ocean, I will know he is there.

The real voyage of discovery lies not in seeking new lands but in seeing with new eyes.—MARCEL PROUST, writer

Dad was never a wealthy man, yet while dying, he said a number of times, "David, I am so rich." In all of our weeks together, he never talked about a set of fancy clothes, a big car, or a special house—that's just stuff.

All we talked about were family, friends, and neighbours, and the things we had done together. That was his wealth—the memories and relationships built over a lifetime—and in those things, he was truly a rich man.

What is the good life? The good life is to be a good neighbor, to consider your neighbor as yourself.—K. VISHWANATHAN, Gandhian activist in Kerala

Dad on the beach at Windy Bay,
Haida Gwaii, 1988

Now that my parents are gone and I too have become an elder, my mind turns to my own mortality. I hope I can approach my death with the dignity and acceptance that my father did.

And I am uplifted by the amazing story emerging from modern science. It tells us that from the moment after the Big Bang, as matter spewed forth in an expanding universe, every particle exerted a pull on every other particle. The universe is not mostly empty space; it is filled with evanescent tendrils of attraction that some believe is the foundation of love, and that attraction is built into the very fabric of the cosmos.

Science informs us that far, far away from the centre of the universe, out in the cosmic boondocks, is an undistinguished galaxy—the Milky Way. And among the billions of stars in that galaxy, our sun is a very ordinary one. And on its third planet—Earth—a mere speck in the heavens, life arose in the last quarter of the cosmos's existence.

And in the very last moment of time, something astonishing happened. A creature emerged from nature endowed with self-awareness, dazzling creativity, and a capacity for love and wonder. Gazing out at a chaotic world, that animal imposed order and meaning in myriad forms and brought humanity to prominence in a cosmic instant. But in a moment of explosive change, we lost our way, forgot the narrative that reminded us of who we are, why we are here, and where we belong.

*Introducing my grandson
to another species*

Tell me the story of the river and the valley and the streams and woodlands and wetlands, of shellfish and finfish. A story of where we are and how we got here and the characters and roles that we play. Tell me a story, a story that will be my story as well as the story of everyone and everything about me, the story that brings us together in a valley community, a story that brings together the human community with every living being in the valley, a story that brings us together under the arc of the great blue sky in the day and the starry heavens at night.—THOMAS BERRY, philosopher

We are the planet's most recent iteration of life's forms, an infant species but one that has the precocity to see our place in the cosmos and dream of worlds yet to come. I believe we are capable of even greater things, to rediscover our home, to find ways to live in balance with the sacred elements, and to create a future rich in the joy, happiness, and meaning that are our real wealth.

I will die before my grandchildren become mature adults and have their own children, but I am filled with hope to imagine their future rich in opportunity, beauty, wonder, and companionship with the rest of Creation. All it takes is the imagination to dream it and the will to make the dream reality.

Acknowledgements

2010 FILM
" FORCE OF NATURE '

THIS BOOK is the result of an unexpected proposal by Laszlo Barna to do a feature film with me. I initially responded with a grandiose proposal based on the story emerging from science about origins—of the universe, the planet, life, and human beings. Wisely, Laszlo and his associates pared it down until, to my surprise, it transmuted into a speech based on the transformation of humanity from just another species to one with geological implications within what I call Living Memory, encompassing the span of my life and that of my elders. Thank you, Laszlo, for making this project possible.

To make the film, Laszlo recruited Sturla Gunnarsson, a renowned director whom I met with considerable trepidation. As I came to know him, my respect for him grew beyond the boundaries of his craft to his intellect and very real humanity. Sturla jettisoned much of the flab of my speech, organized it into an effective discussion, and amplified and added ideas. For the extent to which the speech is effective and powerful, I thank Sturla. As I worked with Sturla, my ideas were shaped into something I'm very proud of and then were made into a film that amazed and delighted me. Thank you, Sturla, my friend.

Laszlo's company included Stephen Silver and Janice Tufford, who made the whole experience of making a film such a pleasure.

Sturla brought the eminent cameraman Tony Westmann into my life, and he contributed ideas and suggestions as well as astonishing pictures.

As well, Kim MacNaughton was always a pleasure to work with, and her strength in toting gear was truly impressive. Rod Matte relieved the tedium of our trips with humour and deadly imitations of people who deserve skewering. Wendy Ord efficiently kept all on track.

Thanks to all the people who made our expeditions happen and the many people who worked on different aspects of the speech.

Turning the contents of the film into a book was the inspiration of Rob Sanders, publisher of Greystone Books. Thanks for insisting, Rob. And the challenge of getting the manuscript finished on time fell on the ever-dependable Nancy Flight.

Neither the film nor the book would have been possible without the indefatigable and indispensable Elois Yaxley.

The long-suffering victims of my absences both physically and mentally were, as always, my family— Tara, Severn, and Sarika. They have never flagged in their support of my efforts and deserve every bit of the credit for all I've done.

Sources for Quotes

*Note: Sources are given in the order in which the quotes
appear in the book.*

Brian Swimme. *The Hidden Heart of the Cosmos*. Video.
1996. Available from the Centre for the Story of the
Universe, Mill Valley, California.

Victor B. Scheffer. *Spire of Form: Glimpses of Evolution*.
Seattle: University of Washington Press, 1983.

Edward O. Wilson. *Biophilia: The Human Bond with Other
Species*. Cambridge: Harvard University Press, 1984.

Zalman Schachter-Shalomi and Ronald S. Miller. *From
Age-ing to Sage-ing: A Profound New Vision of Growing
Older*. New York: Warner Books, 1997.

Loren Eiseley. *The Invisible Pyramid*. New York: Scribner's,
1970.

Aldo Leopold. *A Sand County Almanac*. New York: Oxford
University Press, 1949.

Vladimir Shatalov, quoted in Kevin W. Kelley. *The Home Planet*. London: Queen Anne Press, 1988.

Paul Ehrlich. *The Machinery of Nature*. New York: Simon and Schuster, 1986.

Bernard Campbell. *Human Ecology*. Oxford: Heinemann Educational, 1983.

Susan George. Interview by David Suzuki. Broadcast on British Columbia Knowledge Network, 1997.

John Fowles, quoted in Patrick Curry. *Defending Middle Earth: Tolkien, Myth and Modernity*. New York: Houghton Mifflin, 2004.

Paul Hawken, Amory Lovins, and L. Hunter Lovins. *Natural Capitalism: Creating the Next Industrial Revolution*. New York: Little, Brown, 1999.

Anita Roddick, quoted in Ralph Nader. "The Legacies of Anita Roddick." CommonDreams.org. September 15, 2007. http://www.commondreams.org/archive/2007/09/15/3863.

Clifford Cobb and Ted Halsted. *The Genuine Progress Indicator: Summary of Data and Methodology*. San Francisco: Redefining Progress Institute, 1994.

Robert Kennedy. Speech at the University of Kansas, March 18, 1968. http://www.jfklibrary.org/Historical+Resources/Archives/Reference+Desk/Speeches/RFK/RFKSpeech68Mar18UKansas.htm.

Bernard Lown. "A Prescription for Hope." Nobel Lecture, December 11, 1985. http://nobelprize.org/nobel_prizes/peace/laureates/1985/physicians-lecture.html.

Rachel Carson. *Silent Spring*. Boston: Houghton Mifflin, 1962.

National Religious Partnership for the Environment. *Preserving and Cherishing the Earth: An Appeal for Joint Commitment in Science and Religion*. 1990.

Donald G. Sampson, quoted in David Suzuki and Holly Dressel. *From Naked Ape to Superspecies*. Toronto: Stoddart, 1999.

Johann Wolfgang von Goethe, quoted in Patrick Crean and Penney Kome, eds. *Peace: A Dream Unfolding*. Toronto: Lester & Orpen Dennys, 1986.

Annette Helmer, quoted in Thomas Berger. *Village Journey: The Report of the Alaskan Native Review Commission*. New York: Hill and Wang, 1986.

Harlow Shapley. *Beyond the Observatory*. New York: Scribner's, 1967.

Edward O. Wilson. *The Diversity of Life*. New York: Norton, 1992.

Abraham Maslow. *Motivation and Personality*. New York: Harper and Row, 1970.

Alfred Adler. *Social Interest: A Challenge to Mankind*. New York: Putnam, 1938.

Thomas Berry. *The Dream of the Earth*. San Francisco: Sierra Club Books, 1988.

Marcel Proust. *Remembrance of Things Past*. New York: Vintage Books, 1982.

K. Vishwanathan, quoted in Bill McKibben. "The Enigma of Kerala." *Utne Reader*, March/April 1996.

Thomas Berry. *The Dream of the Earth*. San Francisco: Sierra Club Books, 1988.

References

Note: Sources for references are given in the order in which the references appear in the book. The numbers on the left refer to page numbers.

3   Big Bang, Weinberg, Steven. *The First Three Minutes: A Modern View of the Origin of the Universe.* New York: Basic Books, 1977.

4   Rasmussen, Knud. "Intellectual Culture of the Caribou Eskimos." *Report of the Fifth Thule Expedition 1921–24*, vol. 7, no. 2. Copenhagen: Gyldenhal, 1930.

4   Lévi-Strauss, Claude. *The Savage Mind.* Chicago: University of Chicago Press, 1966.

6   Human beings appeared, Tishkoff, Sarah A., et al. "The Genetic Structure and History of Africans and African Americans." *Science* 324 (2009): 1035–44.

11   Ecological footprint, Wackernagel, Mathis, and William Rees. *Our Ecological Footprint: Reducing Human Impact on Earth.* Gabriola Island, BC: New Society, 1996.

12 Bush, Jeb. On www.cnn.com, November 2001. On December 20, 2006, with a threat of recession, George W. Bush urged Americans, "I encourage you all to go out shopping."

14 Bikini atoll, Ellis, W.S. "A Way of Life Lost: Bikini." *National Geographic,* June 1986.

14 Mueller, Paul. "Dichloro-diphenyl-trichlorethane and Newer Insecticides." Nobel Lecture December 11, 1948. In *Nobel Lectures, Physiology or Medicine 1942–62.* Amsterdam: Elsevier, 1948.

15 Biomagnification, Carson, Rachel. *Silent Spring.* New York: Houghton Mifflin, 1962.

16 Science Summit on World Population. "A Joint Statement by 58 of the World's Scientific Academies." *Population and Development Review* 20 (March 1994).

19 Soil, Nardi, James. *Life in the Soil: A Guide for Naturalists and Gardeners.* Chicago: University of Chicago Press, 2007.

19 Dryland ecosystems, Millennium Ecosystem Assessment. *Ecosystems and Human Well-Being: Desertification Synthesis.* Washington, DC: World Resources Institute, 2005.

20 Atmosphere, Egger, Anne E. "Earth's Atmosphere: Composition and Structure." *Visionlearning* vol. EAS (5). 2003. http://www.visionlearning.com/library/module_viewer. php?mid=107&l=&c3=.

23 Photosynthetic activity, Vitousek, Peter M., Paul R. Ehrlich, Anne H. Ehrlich, and Pamela Matson. "Human Appropriation of the Products of Photosynthesis." *Bioscience* 36 (1986): 368–73.

24 Currents, Broecker, Wallace S. "Thermohaline Circulation, the Achilles Heel of Our Climate System: Will Man-made CO_2 Upset the Current Balance?" *Science* 278 (1997): 1587–88.

24 Biodiversity, NASA Science. "Carbon Cycle." 2010. http://science.nasa.gov/earth-science/oceanography/ ocean-earth-system/ocean-carbon-cycle.

25 Worm, Boris, et al. "Impacts of Biodiversity Loss on Ocean Ecosystem Services." *Science* 314 (2006): 787–90.

28 Fifty thousand species, Toepfer, Klaus. Speech by the Executive Director, United Nations Environmental Programme, at Seventh Conference of the Parties to the Convention on Biological Diversity. Kuala Lumpur, Malaysia, February 2004.

31 Union of Concerned Scientists. "World Scientists' Warning to Humanity." 1992. http://www.ucsusa.org/about/1992-world-scientists.html.

37 Ecosystem services, Wilson, Sara J. *Ontario's Wealth, Canada's Future: Appreciating the Value of the Greenbelt's Eco-services.* Vancouver: David Suzuki Foundation, 2008.

40 Lebow, Victor. "The Real Meaning of Consumer Demand." *Journal of Retailing,* Spring 1955.

42 Simon, Julian. "The State of Humanity: Steadily Improving." *Cato Policy Report,* September/October 1995.

43 Bartlett, Albert A. "The Arithmetic of Growth: Methods of Calculation." *Population and Environment* 14 (1993): 359–87.

46 World Wide Fund for Nature. *Living Planet Report 2006.* Cambridge, UK: Banson Production, 2006.

49 Henderson, Hazel. *Paradigms in Progress: Life beyond Economics.* San Francisco: Berrett-Koehler, 1995.

57 Reimchen, Thomas E. "Salmon Nutrients, Nitrogen Isotopes and Coastal Forests." *Ecoforestry* 16 (2001): 13–17.

63 Unexpected benefits, Bijlefeld, Marjolijn, and Robert L. Burke, Jr. *It Came from Outer Space: Everyday Products and Ideas from the Space Program.* Westport, CT: Greenwood Press, 2003.

66 Guujaaw statement, "Windy Bay." *The Nature of Things.* CBC Television. January 27, 1982.

68 Wilson, Edward O. *Biophilia: The Human Bond with Other Species.* Cambridge, MA: Harvard University Press, 1984.

70 Lungs, Suzuki, David T. *The Sacred Balance: Rediscovering Our Place in Nature*. Toronto: Stoddart, 1997.

70 Shapley, Harlow. *Beyond the Observatory*. New York: Scribner's, 1967.

72 Seventy-one per cent, Keating, Michael. *To the Last Drop: Canada and the World's Water Crisis*. Toronto: Macmillan Canada, 1986.

73 Micro-organisms, Ingham, Elaine. *Soil Biology Primer*. Corvallis, OR: Earthfort, 2010.

75 Human Genome Project Information. 2009. www.ornl.gov/ sci/techresources/Human_Genome/home.shmtl.

77 Scientific contribution, Suzuki, David T. "Temperature-sensitive Mutations in Drosophila melanogaster: Conditional Lethality Is a Useful Feature of Mutations Used in a Variety of Analyses in Higher Organisms." *Science* 170 (1970): 695–706.

78 Harlow, Harry F. "Early Social Deprivation and Later Behavior in the Monkey." In A. Abrams, H.H. Gunner, and J.E.P. Tomal, eds. *Unfinished Tasks in the Behavioral Sciences*, 154–73. Baltimore: Williams and Wilkins, 1964.

79 Institutionalized children, Fisher, Lianne, Elinor W. Ames, Kim Chisholm, and Lynn Savoie. "Problems Reported by Parents of Orphans Adopted to British Columbia." *International Journal of Behavioral Development* 20 (1997): 67–82.

79 Institutionalized children, Nickerson, Colin. "Study on Orphans Sees Benefits in Family Care." *Boston Globe*, November 11, 2006.

80 World Commission on Environment and Development. *Our Common Future*. Oxford and New York: Oxford University Press, 1987.

81 Agroforestry, Halle, Francis. Interview in *The Planet and Us* (Solvay SA), January 2002.

81 Clam gardens, Williams, Judith. *Clam Gardens: Aboriginal Mariculture on Canada's West Coast*. Vancouver: New Star Press, 2007.

83 Faraclas, Nicholas. "Critical Literacy and Control in the
New World Order." In Sandy Muspratt, Alan Luke, and Peter
Freebody, eds. *Constructing Critical Literacies*. Creeskill,
NJ: Hampton Press, 1997.

———{ ∽ }———

About

DAVID SUZUKI

AVID SUZUKI is an award-winning scientist,
environmentalist, broadcaster, and author. He
is renowned for his radio and television pro-
grams, which explain the complexities of the natural
sciences in a compelling, easily understood way, and
he has been the author or co-author of forty-eight
books.

A geneticist, Dr. Suzuki graduated from Amherst
College in 1958 with an Honours BA in Biology and
received a PhD in Zoology from the University of Chicago in 1961. Since 1963, he has been a faculty member of the University of British Columbia, where he is

now Professor Emeritus. He has won numerous academic awards and holds twenty-four honorary degrees in Canada, the United States, and Australia. He was elected to the Royal Society of Canada and the Order of British Columbia and is a Companion of the Order of Canada.

Dr. Suzuki has received consistently high acclaim for his more than forty years of award-winning work in broadcasting. In 1974 he developed and hosted the long-running popular science program *Quirks and Quarks* on CBC Radio. His national television career began with CBC in 1971, when he wrote and hosted *Suzuki on Science*. Since 1979, he has been the host of the award-winning series *The Nature of Things with David Suzuki*. His television series *A Planet for the Taking* won an award from the United Nations, and his BBC/PBS series, *The Secret of Life*, was praised internationally, as was his series *The Brain* for the Discovery Channel.

His many books include *The Sacred Balance* (with Amanda McConnell and Adrienne Mason), *Good News for a Change* and *More Good News* (both with Holly Dressel), *From Naked Ape to Superspecies* (with Holly Dressel), and *Tree* (with Wayne Grady). He is the author of nineteen children's books, including *Salmon Forest* (with Sarah Ellis), *There's a Barnyard in My Bedroom*, and *You Are the Earth* (with Kathy Vanderlinden).

Dr. Suzuki is also recognized as a world leader in sustainable ecology. He the founder of the David Suzuki Foundation, which is a science-based organization dedicated to finding solutions to problems of sustainability, and is the recipient of UNESCO's Kalinga Prize for Science, the United Nations Environment Program Medal, and UNEP's Global 500. In 2009 he won the Right Livelihood Award, which is considered the alternative Nobel.

The David Suzuki Foundation

THE DAVID SUZUKI FOUNDATION works through science and education to protect the diversity of nature and our quality of life, now and for the future.

With a goal of achieving sustainability within a generation, the Foundation collaborates with scientists, business and industry, academia, government and nongovernmental organizations. We seek the best research to provide innovative solutions that will help build a clean, competitive economy that does not threaten the natural services that support all life.

The Foundation is a federally registered independent charity that is supported with the help of over 50,000 individual donors across Canada and around the world.

We invite you to become a member. For more information on how you can support our work, please contact us:

The David Suzuki Foundation
219–2211 West 4th Avenue
Vancouver, BC · Canada V6K 4S2
www.davidsuzuki.org
contact@davidsuzuki.org
Tel: 604-732-4228 · Fax: 604-732-0752

Cheques can be made payable to The David Suzuki Foundation. All donations are tax-deductible.

Canadian charitable registration: (BN) 12775 6716 RR0001

U.S. charitable registration: #94-3204049